Sunset

Basic Carpentry ILLUSTRATED

BY THE EDITORS OF SUNSET BOOKS AND SUNSET MAGAZINE

Lane Publishing Co. ■ Menlo Park, California

Book Editor
Scott Atkinson

Design
Joe di Chiarro

Illustrations
Bill Oetinger

Cover: Photograph by Stephen Marley. Design by Naganuma Design & Direction.

From the ground up . . .

Whether you're a novice carpenter in need of step-by-step instructions, or an experienced do-it-yourselfer looking for a handy reference volume, you'll find what you're after in this complete *Sunset* how-to guide.

Basic Carpentry Illustrated begins with a look at hand and power tools and how to use them. Next, we take on the jumble of building materials, explaining their uses and giving tips on how to select them. The third chapter helps take the mystery out of basic house construction by walking you through the entire building process. And the final chapter presents the "hands-on" information you'll need to build a structure or addition yourself—from the ground up.

In preparing this book, we've relied on many individuals, manufacturers, and organizations for information and advice. We're particularly grateful to L. E. Olstead of L. E. Wentz Co., General Contractors, who shared his knowledge and experience with us. We'd also like to acknowledge the Western Wood Products Association and the American Plywood Association.

Finally, we extend special thanks to Barbara Mathieson for her thorough and thoughtful editing of the manuscript, and to Kathy Oetinger for her many painstaking hours cutting color screens for the illustrations.

Sunset Books
Editor, David E. Clark
Managing Editor, Elizabeth L. Hogan

Second printing November 1986

 Library of Congress Catalog Card Number: 84-80610. ISBN 0-376-01015-0. Lithographed in the United States.

Contents

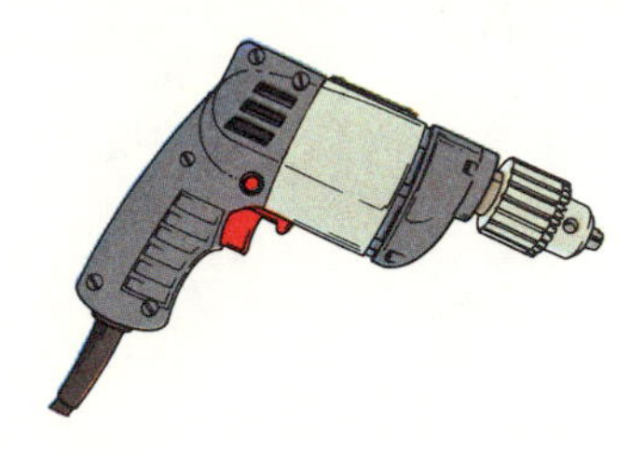

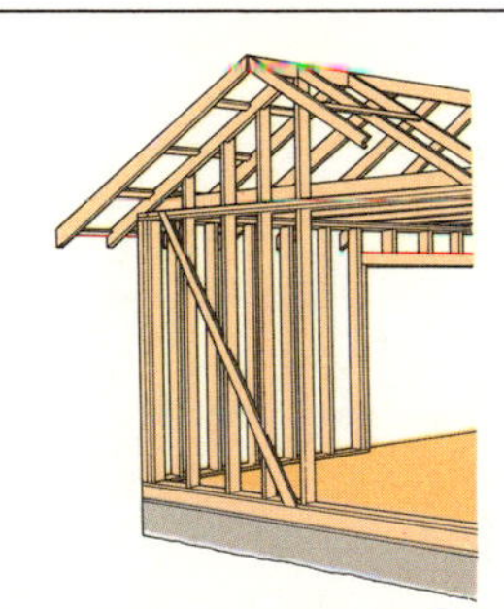

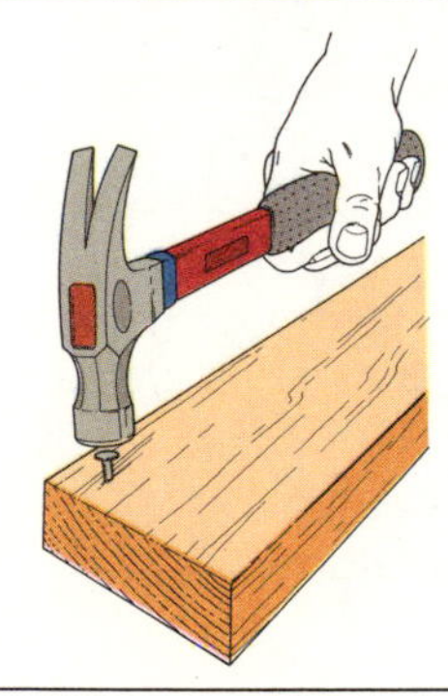

Special Features

Tools & Techniques

Knowing the right tool for the job—and how to use it—is the first step toward becoming a skillful carpenter. "Tools & Techniques" will help you acquire a firm foundation. You can put the chapter to work in two ways: Read consecutively, it provides an overview of the classic carpentry tools and introduces some new tools that fill the gaps in today's do-it-yourself world. But when you're simply looking for the best way to execute a procedure such as measuring, cutting, or drilling, you can turn directly to the appropriate section. We'll help you select the most useful tool by discussing each one's function and design, as well as adjustment and operating techniques.

Working with tools always requires a knowledge of safety procedures; pages 22–23 present a few guidelines and introduce you to some standard equipment that will make your job easier and safer.

Knowledge alone won't produce an ace carpenter, of course. The best way to get acquainted with your tools is to practice on scrap materials. You may also want to look for home improvement classes in your community, or for other opportunities to acquire some "hands-on" experience—*without* making expensive mistakes on your own materials.

Measuring & Marking

The same modest beginnings—accurate measurement, layout, and marking—launch all successful carpentry projects. Though hundreds of tools are available for these crucial first steps, basic carpentry fortunately requires only a small collection. You can go a long way with just a tape measure, a combination square, and a pencil. Add more specialized tools —like a reel tape, carpenter's square, or chalkline—as you need them.

Because the ultimate quality of every project depends on precise dimensions, you'll want to invest in the best measuring and layout tools you can. If you develop careful work habits right from the start, and if you keep in mind the old rule "Measure twice, cut once," you'll be on your way to fine results.

Measuring Tools

To measure distances of a foot or two, you can get by with either a combination square or a rigid bench rule. But for accurate gauging beyond that, a tape measure is generally the answer. A folding wooden rule helps when you need extra rigidity or "inside" measurements. And for laying out distances beyond 25 feet or so, choose a reel tape.

Tape Measure. The flexible steel tape measure is the modern carpenter's workhorse. For all-around utility, choose a 20 or 25-foot tape, 1 inch wide. Be sure it has a locking button to prevent it from retracting—this feature is a big advantage when you're working solo.

A tape measure's end hook should be loosely riveted to adjust for precise "inside" and "outside" readings, as shown below. Many cases are an even 2 or 3 inches in length, aiding inside measurements. Don't trust this, though, until you test it.

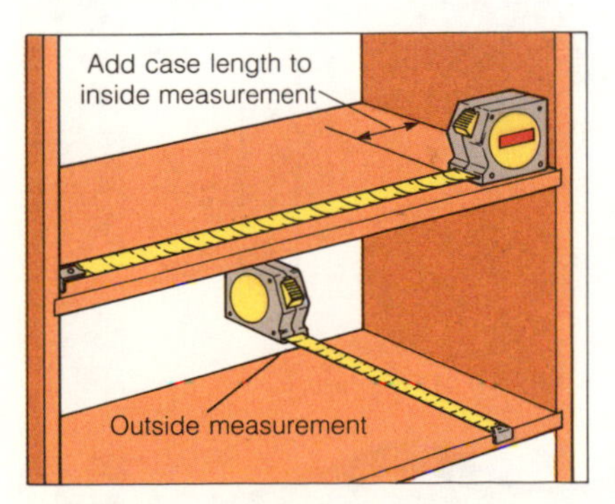

A tape measure's end hook adjusts for inside and outside measurements.

Most tapes are marked in 1⁄16-inch increments. Some smaller models include 1⁄32-inch marks for the first 6 to 12 inches; you'll want one of these (12 feet or so long) if you're doing fine finish work or cabinetry that demands such accuracy. Good tapes also have special marks every 16 inches, the most common spacing for wall studs and floor joists.

■ **Some measuring tips:** Always follow the edge of the material with your tape measure. Pull the tape taut against the end hook before marking the distance. When handling wide sheet materials, measure and mark at several different points.

To measure a wall's height from the floor, butt the end hook against the floor. Pull several feet of tape from the case, and "walk" the slack tape up the wall, bending it at the ceiling. Read the measurement at the bend. Here's where the flexible 1-inch wide tape is handy. These wide tapes will also remain rigid horizontally for 8 to 10 feet if gently pulled from the case—great for measuring openings and distances from corners.

Folding wooden rule. Made from several 6 or 8-inch sections hinged together, the folding rule is generally 6 feet long. Because of its rigidity, the rule extends easily without support at the far end. Some rules feature a sliding extension for accurate inside measurements: add the length of the extension to the length of the unhinged rule body, as shown below.

Reel Tape. When you're laying out the dimensions of a room or foundation, a 50 or 100-foot steel reel tape is a big help. The hinged hook at the end allows you to use the tape solo.

BASIC MEASURING TOOLS

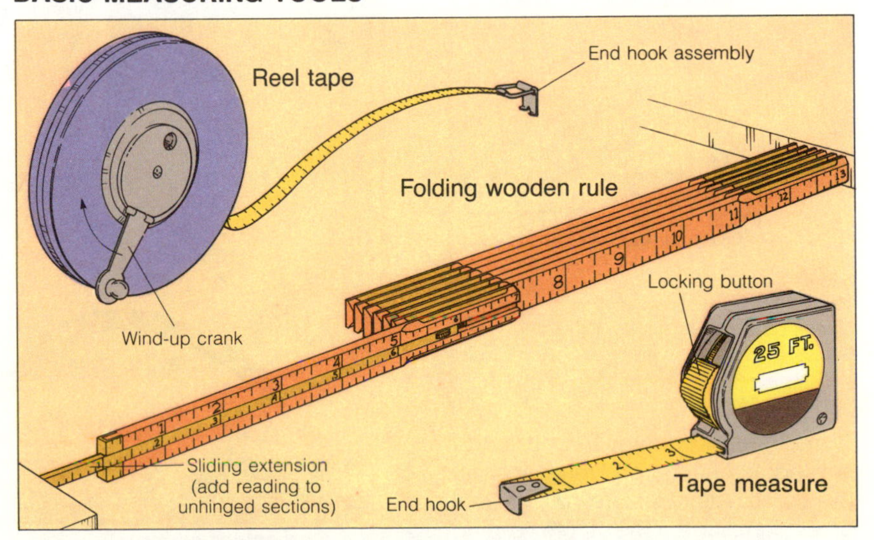

For precise measurements, a 20 to 25-foot tape measure is your workhorse; add a wooden rule and reel tape as you need them.

...Measuring & Marking

Layout Tools

Squares are the primary tools used to lay out and mark cutting lines. Most squares indicate 90° angles; some, in addition, indicate 45° *miter* angles. Adjustable bevel squares can be set to duplicate any angle. Laying out a curve? Your best tool is a French or flexible curve.

Any layout tool must be true, or your most careful work will be wasted. To test a square, hold the body snugly against a straight-edged board and draw a line along the blade. Then flop the square over and draw another line. The two lines should match exactly.

Try square. Squares are typically available with either a 6, 8, or 12-inch blade; the larger ones are best for general work. A try square helps you lay out cutoff lines across boards and framing lumber; some models include a bevel on the handle for laying out short 45° miters.

You can test the squareness of a board's edge or end by positioning the square as shown below. If light shows between blade and board, the board is out of square. To check a board's face, simply lay the blade across the surface.

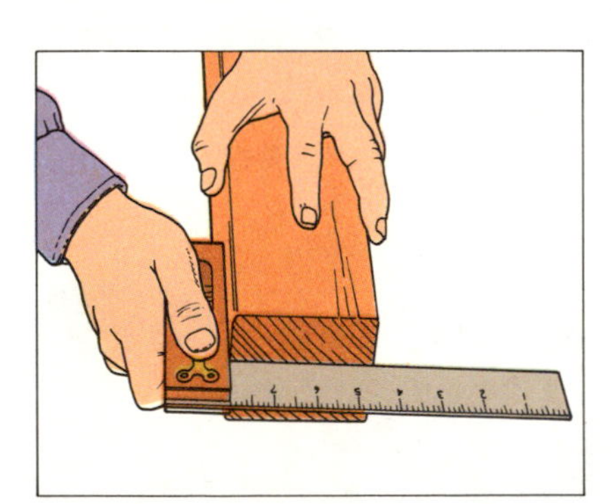

Check board squareness with a try square.

Combination square. In addition to serving as both try and miter square, the combination square has several other functions. The tool's sliding head can be tightened anywhere along the blade, or it can be removed. Many combination squares include a spirit level in the handle for spot-checking level and plumb; the removable scribe on some models is used to mark fine lines.

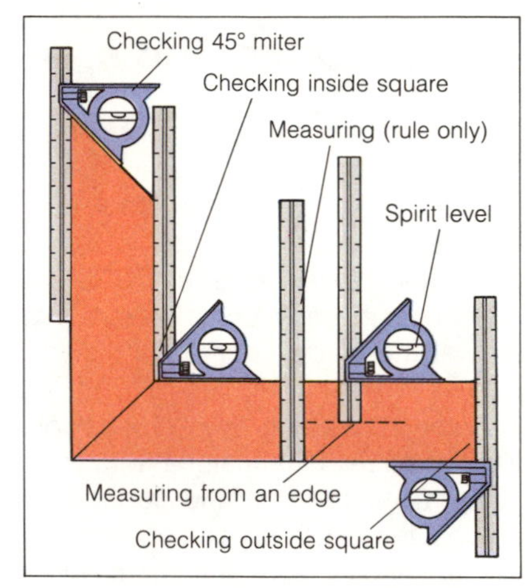

A "combination" of uses makes this tool more versatile than a simple square.

Test a combination square carefully: many modestly priced models have play between head and blade, which can throw the blade out of square. Also be sure to check the blade's increments against your tape measure.

Adjustable T-bevel. A pivoting, locking blade enables you to set the T-bevel at any angle between 0° and 180°. You can determine the correct setting either with the aid of a protractor or simply by matching an existing angle. The latter method is helpful for tying new work—such as a room addition's rafters—into an existing structure.

French curve and flexible curve. Though exasperated carpenters sometimes resort to pie tins and pot lids when laying out curves and arcs, you'll find French curves and flexible curves much handier. Flexible curves are most effective for creating irregular paths between points you've marked. They come in sizes from 12 to 48 inches. Look for both tools at woodworking or drafting supply stores. To lay out full circles, see page 7.

A GALLERY OF LAYOUT TOOLS

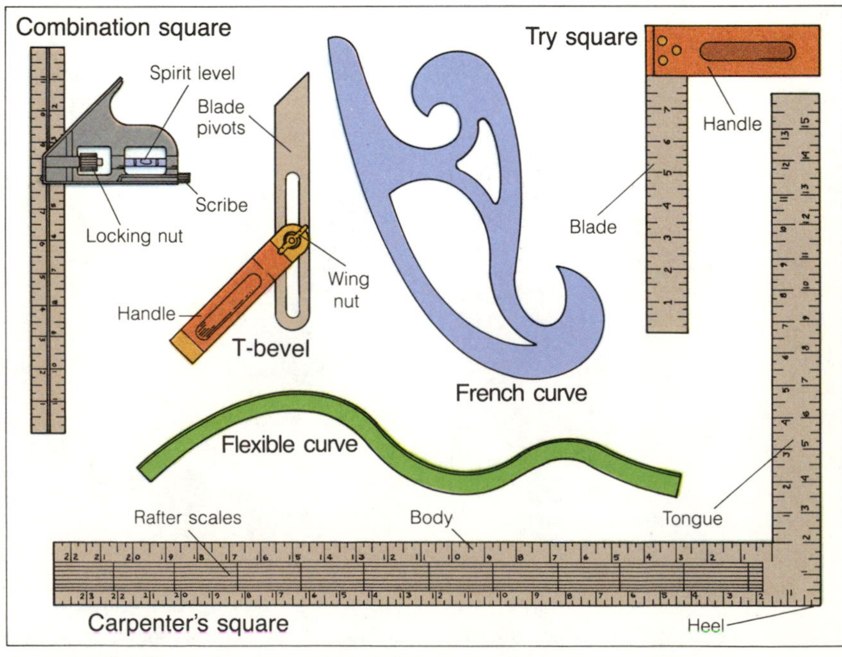

Squares, bevels, and curves help indicate almost any shape of cutting line.

Carpenter's square. When a combination or try square is too small, the carpenter's square is great for laying out lines and checking square. In addition, the figures and tables embossed on the square's face and back enable the carpenter to quickly lay out roof rafters, stairs, or wall bracing.

The tables work on the principle of right triangles: $A^2 + B^2 = C^2$. Once you've determined the lengths of the two straight sides (A and B), the square helps you calculate the length and slope of C as well. For details, see "Ceiling & Roof Framing," pages 75–79, and "Basic Stairways," pages 100–101.

The standard carpenter's square has a 16-inch by 1½-inch tongue and a 24-inch by 2-inch body, meeting at an exact 90° angle at the heel. The most durable squares—and the heaviest—are made from steel. Squares vary widely in the amount of information printed on face and back; those called "rafter" or "framing" squares are normally most complete.

Because the accuracy of the carpenter's square depends entirely on its exact shape, store it where it can't fall or be banged and bent by other tools.

To mark straight lines across a board or sheet, hold the square's tongue against the material's edge and mark along the body. You can also lay out accurate 45° angles by matching the inch gradations on both body and tongue, as shown at right.

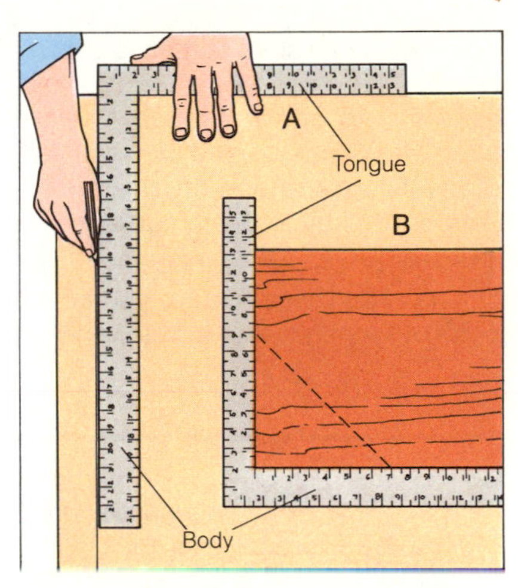

The carpenter's square makes it easy to plot long lines on plywood (A). To lay out 45° angles on sheet materials or solid lumber, match the figures on body and tongue (B).

Marking Tools

Not surprisingly, the trusty pencil serves as your basic marking tool. But choose a hard lead that won't need constant resharpening. Flat-sided carpenter's pencils are durable, stay in place when you put them down, and can be sharpened with a knife to a fine, flat edge. They're awkward, though, for scribing irregular lines or curves.

Mark off distances with a sharp pencil, then carefully recheck each one. When you draw a cutting line, tilt the pencil so that the lead lies flush against the layout tool.

A few additional tools will refine and guide your marking process.

Scribing fine lines. A scratch awl, a utility knife (see page 8), or the scribe on your combination square will mark finer lines than a pencil when precision really counts. Remember, though, that these lines can't be erased in case of error; the pencil line can.

Chalkline. A long, spool-wound cord housed within a case filled with colored chalk, the chalkline is ideal for both marking long cutting lines on sheet materials and laying out reference lines on a wall, ceiling, or floor. For solo work, select a chalkline with an end hook. The type with a pointed case can double as a plumb bob (see page 21).

To mark a line, pull the chalk-covered cord from the case and stretch it taut between two points. Then, toward one end, lift it and release quickly, so that it snaps down sharply—leaving a long, straight line of chalk. For long lines over uneven surfaces, fasten the cord at both ends (it helps to have a partner) and snap it from the center.

Compass or wing dividers. Essential for drawing circles and arcs, a compass or wing dividers can also be used to "step off" equal measurements. Of the two, wing dividers are more precise but cost more; look for the type that allows you to replace one metal leg with a pencil lead.

AIDS FOR MARKING

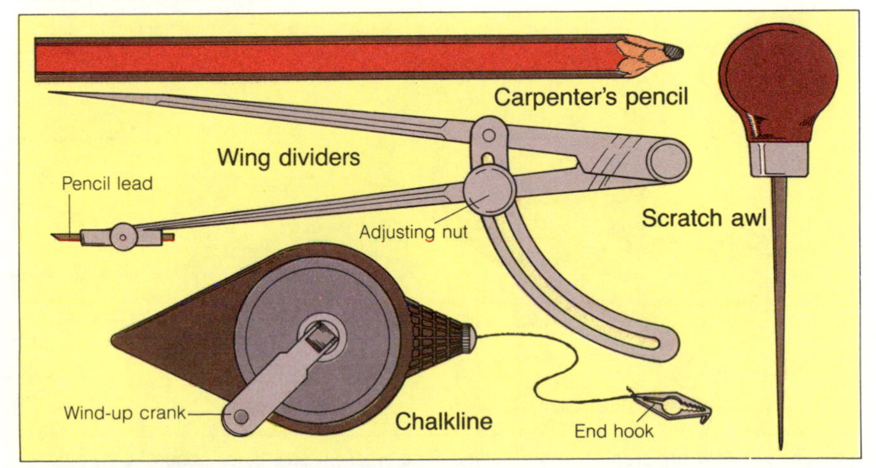

Many marking jobs require only a sharp pencil; scratch awl, chalkline, and wing dividers help refine your work.

Cutting

Accurate, consistent cutting is essential to strong, square joints and assembly. In fact, it's probably no exaggeration to say that a combination of careful cutting and precise measuring is the key to basic carpentry.

Are power saws necessary for cutting? No. If you had to, you could build an entire house with only a crosscut saw, perhaps supplemented by a compass saw. And there are times when only the handsaw will do—spots where a power saw might be dangerous or where no electricity is available. As a rule, though, power saws perform much faster and more accurate work, once you've had some practice.

Two power saws have become standard tools for today's carpenter: the portable circular saw and the saber saw. They are, however, substantially more expensive than most handsaws. Learn to use the basic handsaws first, then move on to the portable power saws as your needs—and wallet—dictate.

Handsaws

A basic collection of handsaws would include the crosscut saw for cutting lumber and sheet products to size, the compass saw for cutouts, and a backsaw and coping saw for finish work and trim. A utility knife, though not exactly a saw, is handy for scoring and cutting gypsum wallboard and other thin materials.

A saw's function is a product of its shape, its blade size, and the position and number of teeth along the blade. The term "point" commonly applied to a saw's blade indicates both tooth size and number of teeth per inch. An 8-point saw has only 7 teeth per inch, since the points at both ends of that inch are included (see drawing at right). In general, the fewer the teeth, the rougher and faster the cut; many teeth means a smooth but slower cut.

Cutting characteristics are also affected by the amount of "set." Saw teeth are bent outward to produce a cut wider than the blade; without this set, the saw would bind in the cut, or "kerf." Once more, the wider set produces a faster but rougher cut. The smaller the set, the finer the kerf.

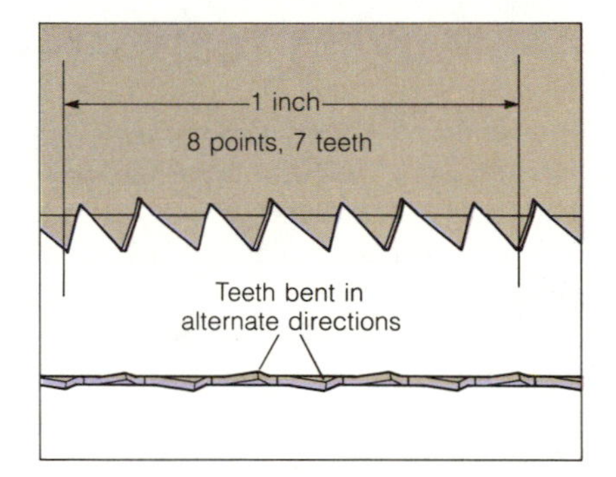

Tooth size (top) and saw set determine cutting performance.

Crosscut saw. Designed to cut across wood grain, the crosscut saw can also be used as an all-purpose saw on plywood and other sheet products. Saw lengths vary from about 20 to 26 inches; the 26-inch blade is a good first choice. For framing work, an 8-point saw is most effective. If you'll need to cut plywood or paneling too, choose the slightly slower—but smoother—10-point saw.

The best crosscut saws are "taper-ground": the thickness of the blade tapers toward the back and the tip. This prevents the saw from binding in the kerf and allows a smaller set to the teeth. Premium woodworker's saws are also "skew-backed," meaning the back is slightly cut away to improve balance and minimize weight. This feature is impractical for an all-around saw you'll work hard.

BASIC HANDSAWS

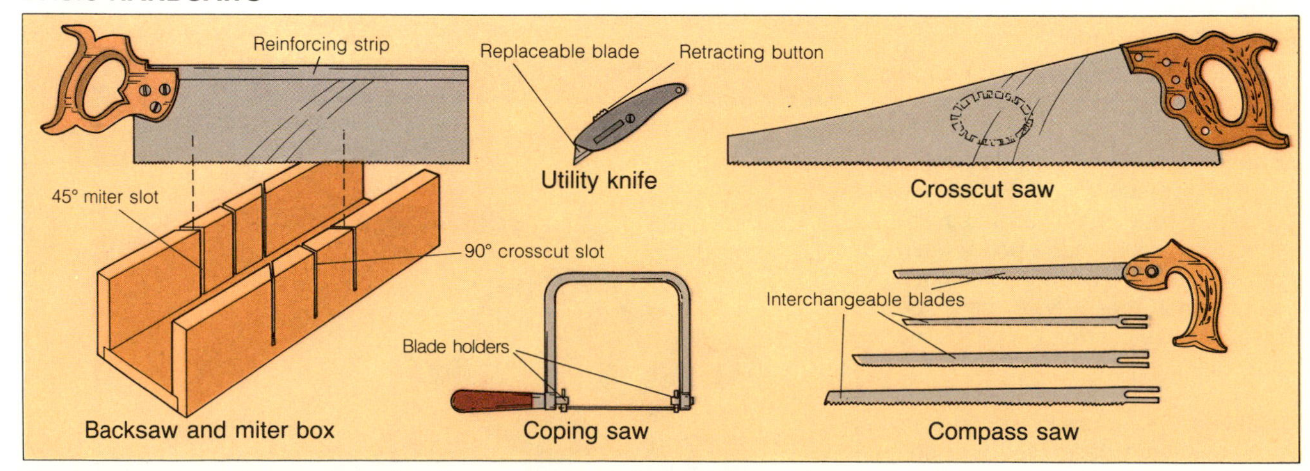

The correct saw ensures accurate cutting. The crosscut saw is for general use; other saws make special cuts. For light-duty cutting and scoring, choose a utility knife.

When you choose a crosscut saw, be sure the handle feels comfortable in your hand. Sight down the saw back to see if it's straight. Flex the tip: it should bounce right back to the center position. High-quality steel, though more costly, will flex better and stay sharp much longer than lower grades.

When your crosscut saw does go dull, the teeth must be filed and (after long, hard use) reset. Take it to a professional sharpener and you'll get a fast, clean job for only a few dollars.

■ **Some crosscutting techniques:** Crosscut saws do about 75 percent of their cutting on the downstroke and 25 percent on the upstroke. Start a cut by holding the saw nearly vertical; slowly draw the blade *up* a few times to make a notch. Use the thumb of your free hand, as shown, to guide the saw at first. A full kerf notched about ½ inch into the far edge of the board will help guide the saw for the remainder of the cut.

If you drew the cutting line right down the middle of your pencil mark (see page 7), be sure all of the saw's kerf is to the waste side of this line—or your finished piece will be too short.

Once a cut is under way, lower the saw's angle to about 45° and progress to smooth, full strokes. Align the saw by sighting down the back from overhead; keep your forearm and shoulder in line with the teeth.

Whenever the blade veers from your cutting line, twist the handle slightly to the opposite side until the blade returns. If you have a persistent problem keeping the blade on target, clamp a straight board along the cutting line to guide the saw.

As you near the end of a cut, reach around the saw, as shown, and support the end of the waste piece. Bring the saw to a vertical position once more and make the last strokes slowly to avoid breaking off—and splintering—the board. For long plywood cuts, recruit a helper to hold the waste piece.

Ripsaws. A specialized version of the crosscut saw, the ripsaw has larger, chisel-like teeth that cut rapidly in line with the wood grain. Ripsaws generally are available with 5, 5½, or 6 points per inch. If you're planning to buy a portable power saw, skip the ripsaw—your crosscut saw will handle any occasional hand-ripping tasks.

Operate the ripsaw in the same manner as a crosscut saw, but hold the blade at a slightly steeper angle—about 60°.

Compass (keyhole) saw. No hand tool surpasses the compass saw in making cutouts and gentle curves. The thin blade, 10 to 14 inches long, tapers from about an inch at the handle to a pointed tip. Typical blades have 8 points per inch, producing a fast but relatively rough cut. A smaller version, 6 to 8 inches long with a straight handle, is designed for making cutouts in gypsum wallboard.

To begin a cutout in wood, first drill a pilot hole for the compass saw's blade. After the cut is started, you can switch to a crosscut saw for a long, straight cut.

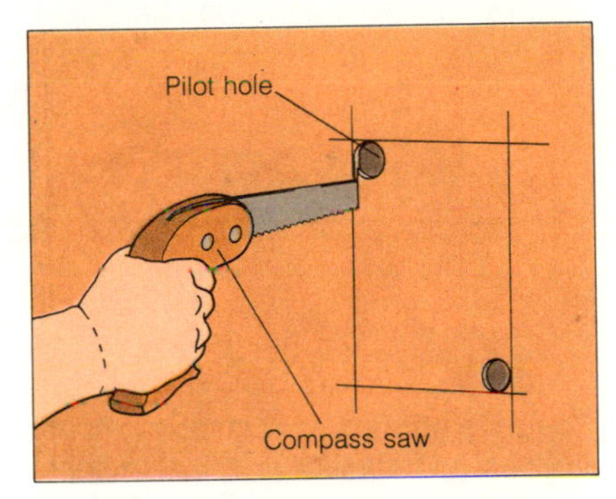

For cutouts in wood, insert a compass saw in predrilled pilot holes.

A cutout in wallboard doesn't require a pilot hole: simply tap on the saw's handle end with your free hand or a hammer until the blade is started.

CROSSCUTTING TECHNIQUES

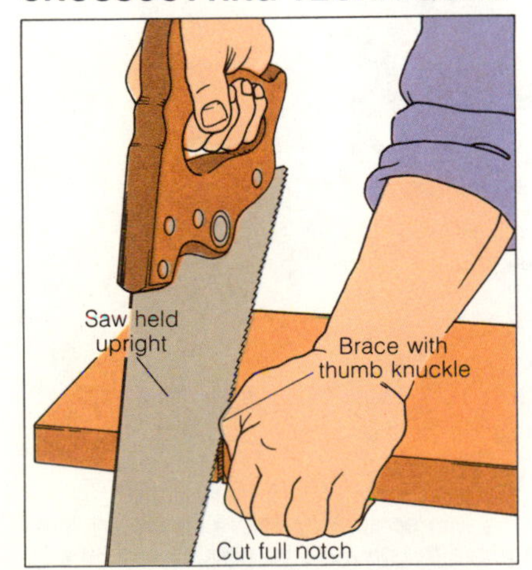

To start a cut, hold the saw upright and guide the blade with your thumb knuckle. Make a full notch in the board's end.

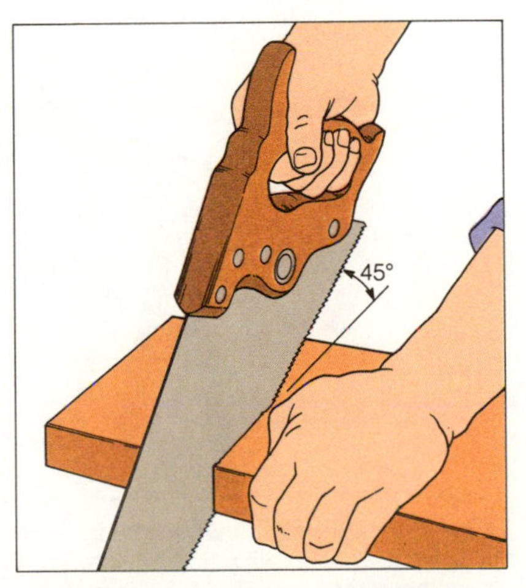

Once the cut is underway, lower the angle to 45° and cut with full, even strokes.

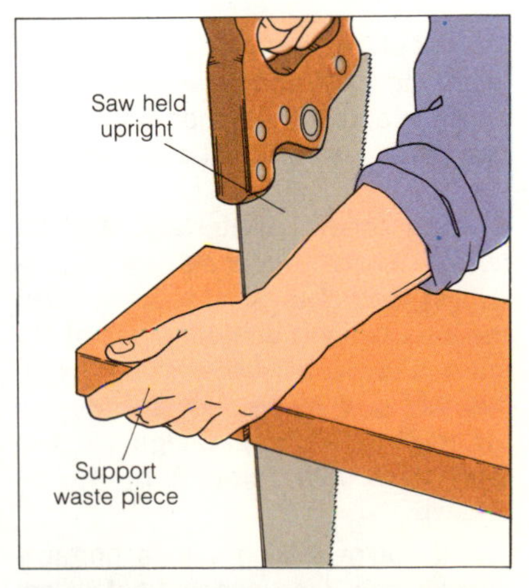

At the end of a cut, support the waste piece with your free hand; finish with short, upright strokes.

. . . Cutting

Backsaws. Designed for crosscutting fine finish work like moldings and trim, the rectangular backsaw derives its name from a metal reinforcing strip that runs the length of the back. The reinforcement prevents bowing and allows use of a thinner blade, producing very fine, straight cuts.

The typical backsaw has a 12 to 14-inch blade with 12 to 16 teeth per inch. The 12-inch saw with 12 tpi is a good first choice.

A miter box (see drawing, page 8) is often used to guide the saw into materials at a fixed 90° or 45° angle. Integral backsaw/miter box units that cut to any angle are also available; the saw in these units ranges up to 26 inches in length. Though more versatile and precise, these units are quite expensive. *Power miter saws* are handier yet; if you have a lot of finish work to do, consider renting one.

Unlike the blades of the crosscut saw and ripsaw, the backsaw's blade is held parallel to the work surface.

Coping saw. A thin, wiry blade strung taut within a small, rectangular frame enables the coping saw to make fine, accurate cuts and follow tight curves—but cutting is limited to surfaces that its relatively shallow frame will fit around. Typical throat depth is 4¾ inches, but some manufacturers offer deeper models. Coping saw blades average 6½ inches in length; teeth per inch range from 12 to 15.

Blades may be positioned with teeth up, down, or to either side; rotate the blade holders (see drawing, page 8) to adjust the position.

For cutouts near an edge, slip the blade through a pilot hole and then reattach it to the frame. Clamp the material to a sawhorse or vise for better control. If you use a sawhorse, point the teeth away from the handle and cut on the push stroke. But when cutting through material held in a vise, point the teeth toward the handle and cut on the pull stroke.

SUPPORT—FOR CLEAN, SAFE CUTTING

To make efficient saw cuts, you must support your lumber or sheet material securely. In a well-equipped shop you'll have no problem, but if you must go to a jobsite, you'll need a pair of sturdy yet portable sawhorses, or a folding, portable workbench.

In sawhorses, you have two options: build your own from scratch, or add 2 by 4 or 2 by 6 crossbraces to purchased metal folding legs.

Crosscutting lumber. When cutting lumber to length, you may need only a single support. For a short cut, simply hang the waste end off one edge.

Cutting sheet materials. Cutting across sheet materials (or crosscutting long boards near the middle) requires support on both sides of the cut so the waste neither tilts in (binding the saw blade) nor swings out (splintering the cut). Bridge the two sawhorses with scrap 2 by 4s, as shown.

If you're cutting with a portable circular saw (see pages 11–12), set the blade depth so that you cut through the material but just nick the scrap. With a handsaw, cut only up to the scrap support. Slide the sheet forward slightly for a little more unobstructed cutting; then reposition the sheet and continue cutting on the other side of the support. Ideally, you'll have a helper on hand to hold the waste piece.

Usually, ripping only requires two sawhorses and no scrap supports. (Thin sheets may need support below the cutting line to prevent sagging.)

SAWHORSE SETUP

To evenly support plywood, bridge sawhorses with scrap 2 by 4s. When cutting with a handsaw (inset), saw up to the scrap; then move the plywood as needed and saw past the scrap.

Portable Circular Saw

Using a circular saw and the correct blade, you can easily cut 10 times as fast as with the crosscut saw. Circular saws are commonly made in sizes from 5½ to 8¼ inches; the size refers to the largest diameter of blade that fits the saw's arbor (axle). The popular 7¼-inch saw cuts through surfaced 2-by framing lumber (see page 35) at any angle from 45° to 90°.

Two distinct styles of circular saw are available: the standard, or "sidewinder," and the worm-drive. For heavy use, the more expensive worm-drive offers durability as well as a motor housing that sits behind the blade, creating better balance and allowing the right-hander an unobstructed view of both blade and cutting line.

Whichever type you select, your saw should be equipped with both depth and angle adjustment levers; an upper, fixed blade guard; and a lower, spring-action blade guard (see drawing above right). One handy accessory, the ripping fence, helps guide straight cuts near the edge of a board or sheet.

ELEMENTS OF A CIRCULAR SAW

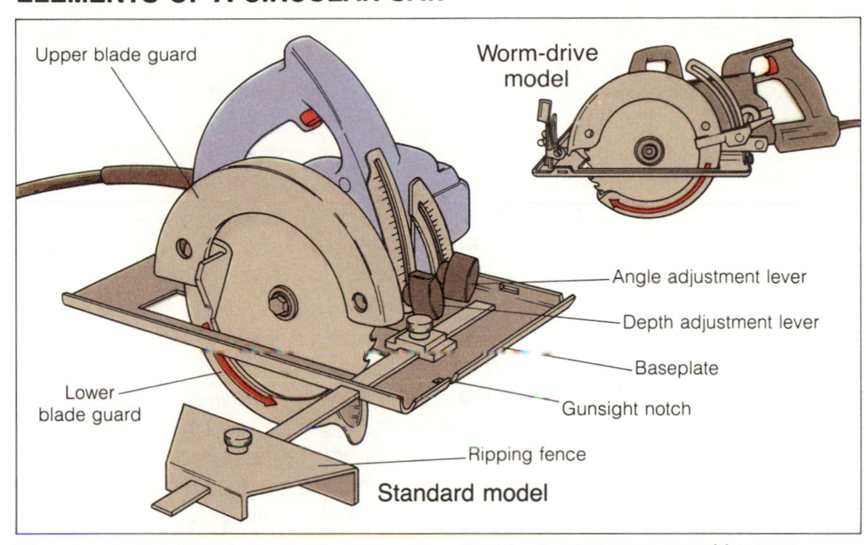

The standard "sidewinder" saw is most practical for beginners; worm-drive saw lets you see the cutting line.

Choosing the right blade. Be sure you have the correct size and type of blade before you start work. Though terms often vary among manufacturers, these are the most common types:

The *combination* blade normally comes with your saw. Combination blades unite rip and crosscut principles into a single cutting edge suitable for everyday needs, though sacrificing a bit of the precision and speed of a specialized blade. *Planer* blades are the finest-cutting combination blades.

Cutoff blades have finer teeth than combination blades for clean crosscutting. *Plywood* blades are finer still and are specially tempered to resist the abrasion of plywood glues. *Rip* blades have wide, chisel-like teeth for ripping boards with the grain.

Carbide-tipped blades, though substantially more expensive than standard steel blades, last many times longer. When finally dull, they (like any circular saw blade) can be resharpened for a few dollars.

Cutoff wheels are abrasive discs designed to cut a variety of metal and masonry materials.

CIRCULAR SAW BLADES: A BASIC SELECTION

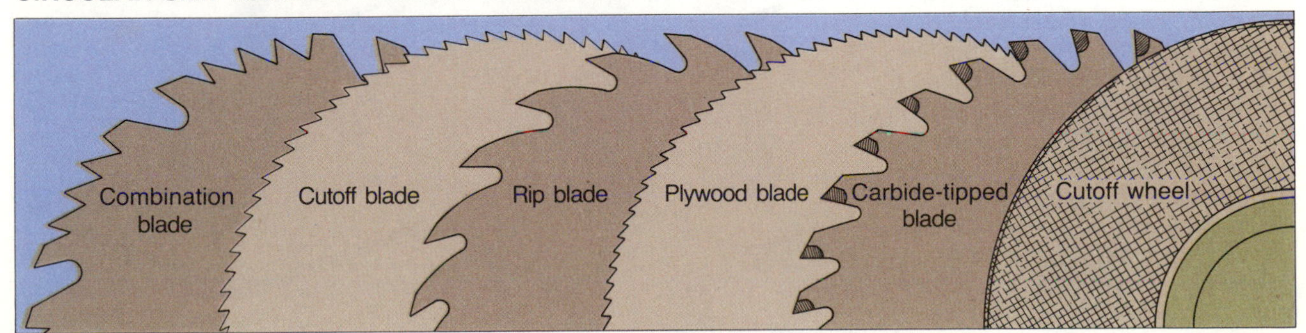

Choose a blade to fit your task. The combination blade is a jack-of-all-trades; cutoff, rip, and plywood blades produce finer cuts. Carbide-tipped blades stay sharp longer; the cutoff wheel takes on masonry and ceramics.

. . . Cutting

Saw setup. Before you start that first cut, check blade type, blade depth, and the cutting angle. And don't make any adjustments on your saw until you're certain that it's *unplugged*.

First, attach the correct blade for the job. To change blades, you'll need a wrench that fits the saw's arbor nut. You'll also need to stop the blade from rotating: either lightly dig the teeth into some scrap lumber or use the button on some models that freezes the blade.

The arbor nut is "reverse-threaded"—it comes off clockwise and tightens counterclockwise. When installing the new blade, be sure that the teeth point forward and up (see drawing, page 11).

For most cuts, you'll want the blade adjusted to 90°. Loosen the angle adjustment lever and push on the baseplate until it stops in the horizontal position; tighten the lever. If you're cutting at an angle (beveling), set the blade with the help of the degree scale on the saw body. But don't trust the accuracy of this setting until you've tested it on a scrap.

Next, loosen the depth adjustment lever and set the correct blade depth. Either measure this with your tape measure or place the baseplate on the material to be cut and "eyeball" the depth. For most materials, you'll want the blade to protrude only 1⁄16 to 1⁄8 inch below the cut. When crosscutting standard 2-by framing lumber, though, leave the blade at maximum depth.

Basic operation. Because the blade cuts in an upward direction, the material's top surface tends to splinter; place the best side down. To start a cut, rest the saw's baseplate on the material and line up the blade with the waste side of your cutting line. Don't let the blade touch the material yet. Be sure your power cord or extension is free of the cutting path and won't get tangled in the work as you progress. And be sure to wear safety goggles (see page 22).

Release the safety button, if you have one. Let the motor reach speed; then slowly feed the saw into the material. Depending on your saw type, you'll either aim the blade directly from the side or use the gunsight notch on the baseplate. (Gunsight notches are often inaccurate—be sure to test yours.)

If the saw binds, make certain that your support is adequate. Back the saw off about an inch and try again. On long cuts, a large nail or screwdriver placed in the kerf helps prevent binding.

As you reach the end of the cut, be sure you're in position to support the saw's weight. If necessary, grip the front handle with your free hand. When you're crosscutting an unsupported piece, like a 2 by 4 stud, accelerate right at the end to avoid splintering. And always let the blade stop completely before you swing the saw up or set it down.

Ripping. To rip a board to width, set the blade at the minimum depth and attach the ripping fence loosely. Lining the blade up with the correct width marked at the board's end, tighten the screw holding the fence. Be sure to account for the blade's kerf.

Ripping can be slow, dusty work and is especially prone to kickback. Cut by pushing the saw slowly away from you. When you need to reposition yourself, back the saw off an inch or so in the kerf and let the blade stop while you move farther down the line.

Saw guides. What's the secret to really straight cuts with the circular saw? Clamp a straight length of scrap lumber to the material to guide the baseplate of the saw (see page 27 for details on clamps). Measure from the blade to the edge of the baseplate; clamp the guide at that distance from your cutting line.

Special problems. Two awkward cutting situations occasionally crop up: cutting a very narrow piece from the edge or end of a board or sheet, and cutting a post or beam that's thicker than the saw's capacity.

Narrow pieces don't provide enough support for the saw's baseplate; try butting a short piece of scrap against the edge or end of the board. If the blade guard jams due to lack of resistance, lift it manually by the lever. Never tie or jam open the guard.

To cut oversize lumber like 4 by 4 posts, first extend the cutoff line around all four sides with your try or combination square. Set the saw blade to the maximum depth, and cut through one side. Then flip the piece over and cut through the back. Smooth any unevenness with a block plane, rasp, or sander.

TWO COMMON CUTS WITH A CIRCULAR SAW

When ripping a board to width, set blade depth just below the board and guide the cut with a ripping fence.

For long, straight cuts, clamp a scrap guide to the material for the saw's baseplate to ride against.

Portable Saber Saw

The saber saw's high-speed motor drives one of many interchangeable blades in an up-and-down (reciprocating) motion. This saw's specialty is curves, circles, and cutouts, but you can also use it for straight cutting or beveling.

Saber saws are available in single-speed, two-speed, and variable-speed models. A variable-speed saw accelerates as you squeeze down on the trigger, allowing fine control when cutting tight curves or different materials. Look for a saw with a tilting baseplate for cutting bevels to 45°. An adjustable ripping fence helps guide straight cuts parallel to the edge of a board or sheet.

Choosing the right blade. In general, blades with 4 to 7 teeth per inch are designed for rough, fast cuts in wood. Fine finish work, tight curves, and scrollwork require blades in the 10 to 20 tpi range. Metals demand even finer teeth—24 to 32 tpi.

A variety of specialty blades cut plastics, rubber, or leather; blades embedded with carbide chips even tackle ceramics. A flush-cutting blade allows you to cut forward up to a wall or another obstruction.

Be sure the tang (the shank) of any blade you choose fits your saw's locking device.

Basic operation. Since the saber saw's upward-cutting blade may cause the material's top surface to splinter, you'll want to place the best side down. Be sure to wear safety goggles.

Guide the saw through straight cuts as you would the circular saw (see page 12). Cut slowly: fast cutting leads to snapped blades and an overheated motor. For really straight cuts, you'll need a guide for the baseplate.

When cutting curves, use the thinnest blade you can. And remember: the tighter the curve, the more slowly you should cut.

ELEMENTS OF A SABER SAW

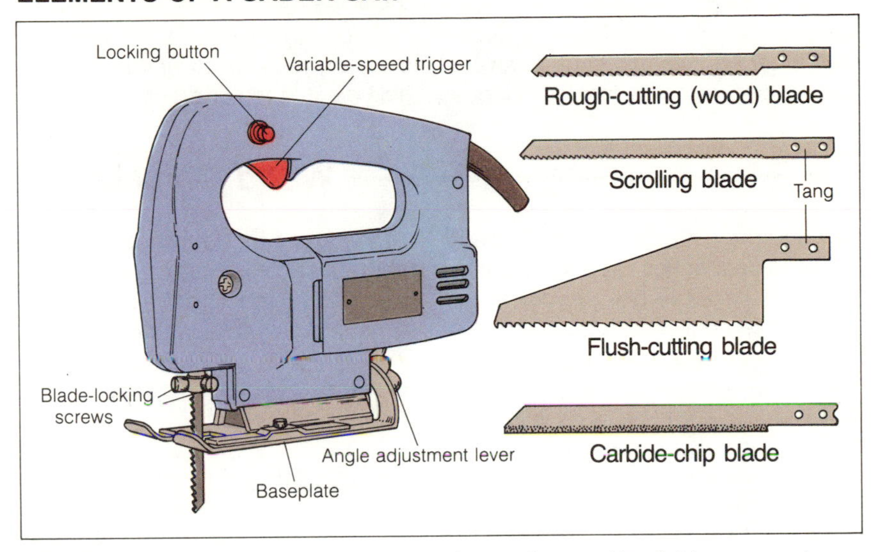

The saber saw, equipped with the proper blade, excels at making tight curves and cutouts.

Cutouts. To cut circles or rectangular cutouts inside a panel, it's simplest to first drill a pilot hole for the blade in the waste area.

With practice, you can also start the cutout in thin, soft materials by "plunge-cutting" with a rough-cutting blade. Rock the saw forward onto the baseplate's front edge until the blade is free of the material. (Remember: the blade moves up and down—be sure it will clear the surface at its *longest* point.) Turn the saw on; then slowly lower it until the blade tip cuts into the material and the baseplate rests flat on the surface.

Circles with a radius as large as 6 or 7 inches can be executed with a circle guide (often the ripping fence turned upside down).

TWO SPECIAL CUTS WITH A SABER SAW

To plunge cut, tilt the saw, turn the motor on, then slowly lower the baseplate onto the material.

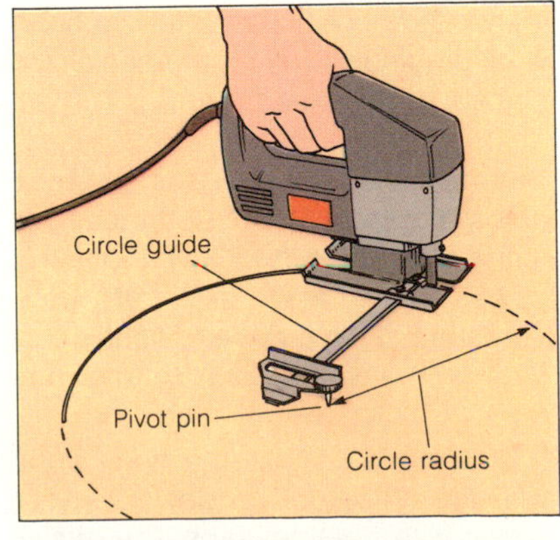

To use a circle guide, align the pivot pin with the front of the blade; length from pin to blade determines the radius.

Shaping

Many carpentry projects require additional cutting, shaping, and smoothing before pieces can be assembled.

Chisels perform a variety of tasks —rough shaping of framing members, paring notches and grooves, and cutting mortises for door hinges and hardware. Jobs such as smoothing surfaces, squaring boards, and making fine joint adjustments call for a plane.

For sophisticated joinery and shaping, you'll discover that the portable electric router makes short work of many painstaking tasks formerly accomplished with chisels and specialty planes.

Turn to files, rasps, and perforated rasps when shaping and smoothing wood or metal to its final form.

Wood Chisels

Although a browse through woodworking catalogues will turn up a labyrinth of chisel styles, the carpenter can concentrate on three basic types:

Bench (bevel-edged) chisels, with 4 to 6-inch-long blades, have side bevels to fit tight spots. A shorter version, the *butt chisel,* serves frequently for all-purpose cutting. Blade widths typically range from 1/4 inch to 2 inches.

Firmer (framing) chisels have squared-off sides and longer blades—up to 11 inches long—for the heavy-duty, deep cutting and paring needed for post-and-beam or timber framing (see pages 60–61). Blade widths vary from ½ inch to 2 inches.

Mortise chisels—long, narrow, and square-edged—are designed for carving out deep recesses. Typical blade widths run from ¼ to ½ inch.

Most carpenters choose one or two plastic-handled, steel-capped butt chisels for rough work; the capped handles can be driven with a hammer. Blade widths of ¾ and 1½ inches are good choices. For finish work or joinery, buy more specialized chisels as you need them.

Shaping a notch. To easily chisel a notch or groove, first cut the outlines to the proper depth with a handsaw or power saw—ideally, a circular saw. Then make several additional cuts through the waste area, as shown at right.

Remove any remaining wood with the chisel, bevel down. Drive it lightly with hammer taps; then finish smoothing the bottom with hand pressure alone, bevel up.

Shaping a hinge mortise. To chisel the recess for a hinge or other hardware, first trace its outline. Score the lines, first with a sharp knife and then with light taps on a bench or mortise chisel. Always keep the chisel's bevel facing waste wood.

Make a series of parallel crossgrain cuts to the proper depth, as shown. Lower the angle of the chisel; using hand pressure, chip out the waste wood. Smooth from the side, if possible, holding the chisel almost flat, bevel up.

BASIC CHISEL TYPES

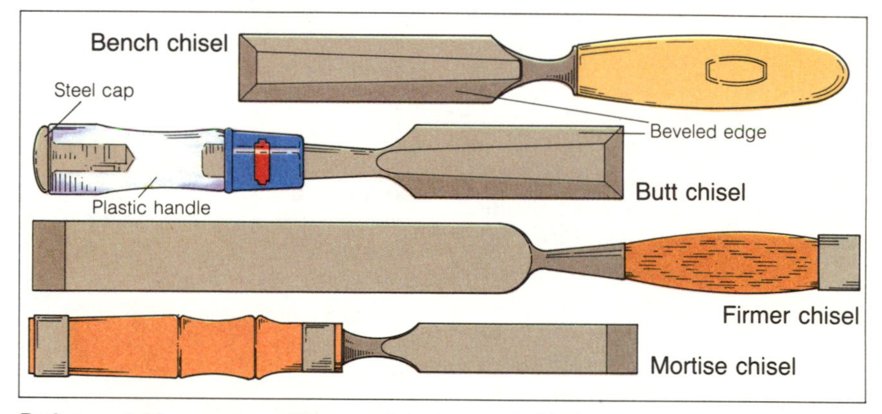

Paring, notching, and mortising are jobs for wood chisels.

SHAPING A NOTCH

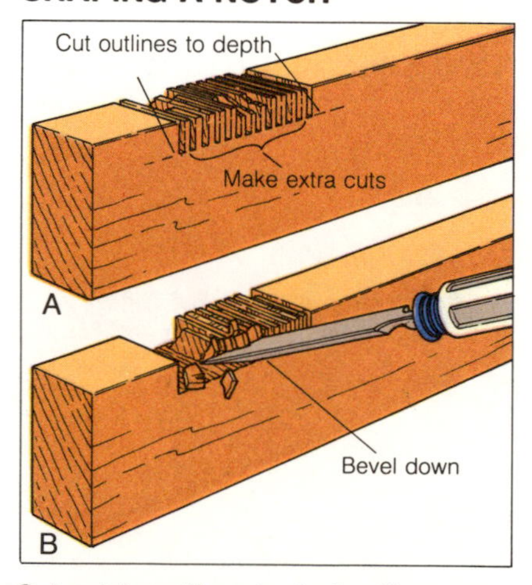

Cut notch outlines to depth with a saw, then make extra cuts through the waste (A); chip waste out with a chisel (B).

SHAPING A HINGE MORTISE

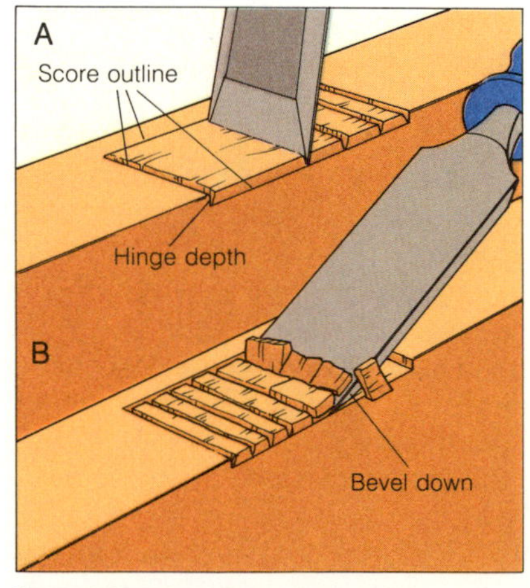

Score hinge outlines, then make a series of parallel chisel cuts (A); chip out waste wood, bevel down (B).

Planes

Planes that carpenters commonly use fall into two basic categories: *bench* and *block* planes. Bench planes smooth and square in line with the grain. Three main types are the jointer plane (about 22 inches long with a 2⅜-inch-wide blade), the versatile and popular jack plane (14 inches by 2 inches), and the smooth plane (9¾ inches by 2 inches).

The shorter block planes (typically 6 inches by 1⅝ inches) smooth end grain, cut bevels, and trim small bits of material from boards that don't fit snugly.

Adjusting a plane. To get the most from your plane, you should understand its components and keep them in fine adjustment.

A lever cap holds the bench plane's cutting iron and cap iron under tension against the sloped body, or "frog." Your first adjustment should be setting the clearance between cutting iron blade and cap iron. Lift up the locking lever; then remove the lever cap to free the two irons. Loosen the screw from below the cutting iron and adjust the alignment between irons—about 1⁄16 inch of the cutting iron's blade should be exposed. Tighten the screw; then reassemble.

Check the angle and exposure of the blade by turning the plane over and sighting down its sole. If the blade is out of square, push the lateral adjustment lever toward the side that's extended farthest. To adjust blade exposure, turn the depth adjustment nut until the blade just protrudes through the mouth.

Block plane adjustments vary according to the model. The "fully adjustable" type has both lateral adjustment lever and depth adjustment nut.

"Adjustable" block planes may include a depth adjustment nut and/or a locking lever for the cutting iron assembly. On these models, you'll need to loosen the locking lever and adjust blade angle (and possibly blade depth) by hand.

Operating tips. When working with a bench plane, grip the rear handle with one hand, the front knob with the other. Slightly angle the plane so it makes a shearing cut. Always cut in the direction of the grain. For clean, shallow cuts, determine how the grain slopes and cut "uphill," as shown below.

Hold a block plane in one hand, as shown below, applying pressure to the front knob with your forefinger as necessary. Use short, shearing strokes to cut end grain. To prevent splitting a board's edge when planing end grain, plane inward, slightly bevel the edge first, or clamp a piece of scrap wood to the edge.

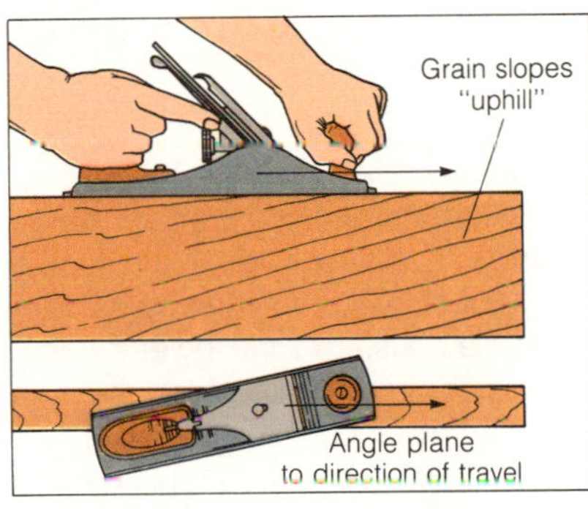

Guide a bench plane with both hands and cut "uphill" with the grain.

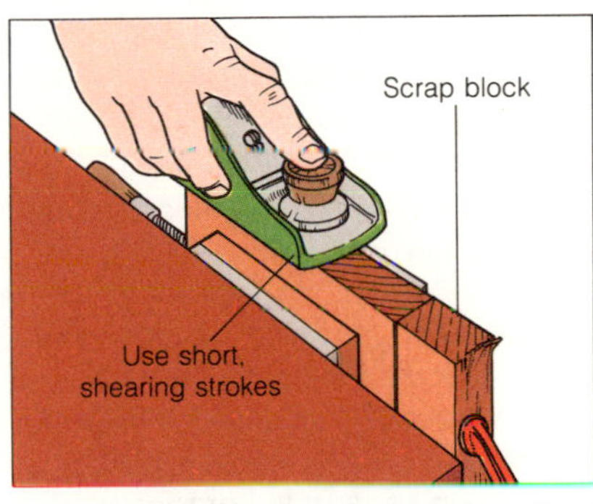

A scrap block keeps the block plane from breaking away a board's edge.

ELEMENTS OF BASIC PLANES

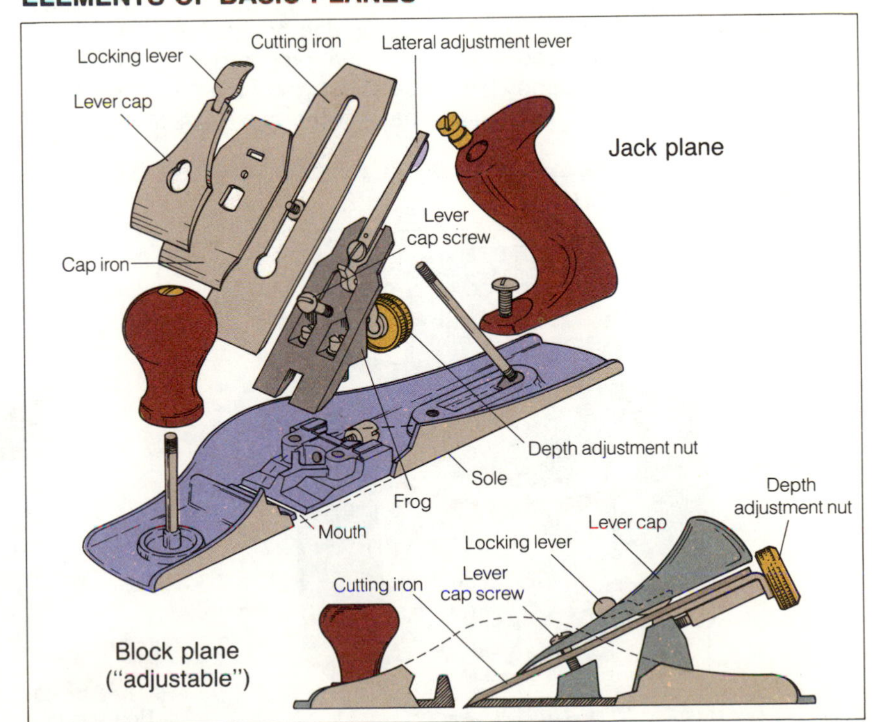

Jack and block planes are a carpenter's main planes. The medium-size jack plane (top) smooths and squares boards; choose the block plane (bottom) for small jobs like end grain.

. . . Shaping

Electric Router

A router equipped with the proper bit cuts all kinds of grooves: dadoes, V-grooves, rounded grooves, and even exact dovetails. It can also round or bevel the edges of a board, trim plastic laminate at a single pass, and whisk out hinge mortises in minutes with the aid of a template.

Essentially a stand-up motor that rotates a bit at speeds up to 25,000 rpm, the router produces fast, clean cuts. The motor and bit are normally raised or lowered within a protective housing to adjust cutting depth.

Desirable features include two comfortable hand grips at opposite sides of the housing, a trigger that's incorporated into one of those handles (safer than a toggle switch on the motor), a trigger-activated light, and a convenient depth-adjusting ring or knob with accurate gradations. It pays to spend a little extra for a router with at least one horsepower; otherwise, the depth of cut possible in one pass is very limited.

Choosing router bits. For normal use, bits made from high-speed steel are sufficient. Carbide-tipped bits cost more but stay sharp longer—choose them for hard woods, particleboard, plastic laminate, or synthetic marble.

The most popular bits are shown below. Edge-cutting bits typically combine with self-piloting tips. Some bits include the pilot; for others, you must purchase the pilot and arbor (shank) separately. A guide bushing, screwed to the router's baseplate, keeps a mortising bit from cutting the template.

Operating the router. To set bit depth, first place the router's baseplate on a flat surface. Loosen the wing nut or knob that clamps the motor to the housing. Then lower the motor until the router bit just touches the flat surface. (When adjusting an edge-cutting bit, let the pilot hang over an edge, and set the bit's cutters on the flat surface.)

The exact mechanism for setting bit depth varies from one router to another. You generally set the depth indicator to zero and then turn the adjusting ring or knob until it's lined up with the proper depth. For a very deep cut, you'll need to make a first pass at the maximum depth on the indicator. Then reset the indicator to zero, readjust the ring or knob for the remainder, and make a second pass. Happily, on any router you can also ensure an accurate setting by simply measuring the clearance between bit and baseplate, then testing the depth on a piece of scrap.

Because the router bit spins in a clockwise direction, it tends to drift or kick back counterclockwise in your hands. To compensate, you'll normally operate the router from left to right, so the cutter's leading edge bites into new wood.

Gripping the router securely by the handles, line it up just outside the area to be cut. Let the motor gain full speed before starting to cut; then carefully feed the bit into the work. Be sure to wear safety goggles.

It takes some looking and listening to acquire a feel for the correct speed. Listen for the sound of the router's motor that corresponds to the smoothest cut, and you'll have little trouble.

At the end of a cut, turn the motor off as soon as the bit is clear; let the bit stop completely before setting the router down.

ELEMENTS OF A ROUTER

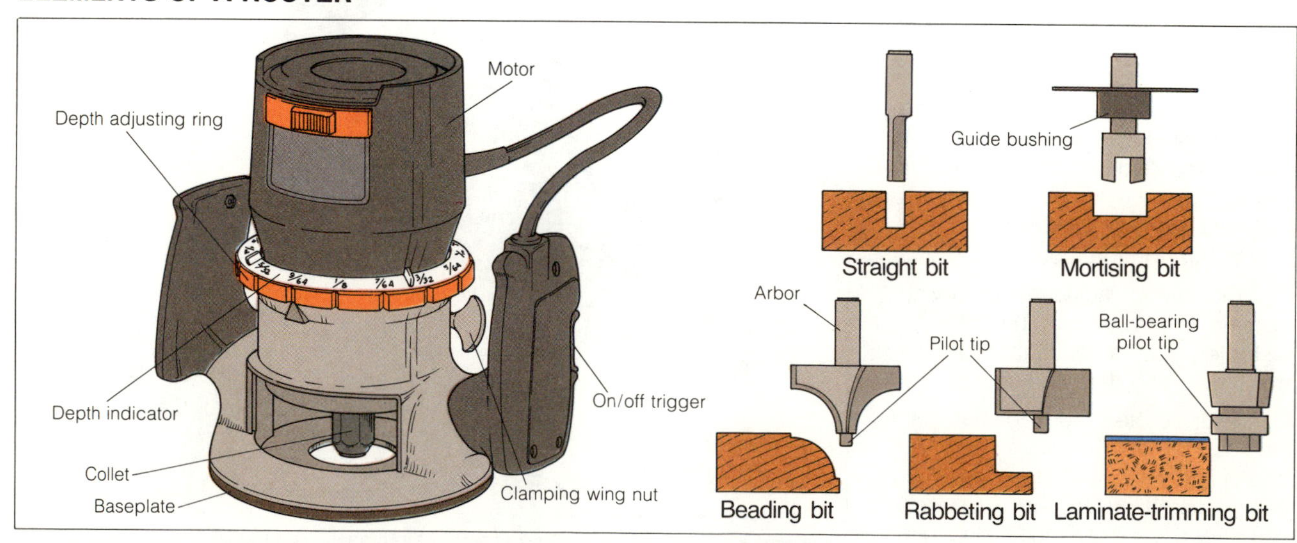

The electric router makes short work of grooves, notches, and edge trimming when fitted with the correct bit.

THREE COMMON ROUTER CUTS

To rout out a groove (dado), set up a guide for the router's baseplate; cut with a straight bit.

To shape a rabbet or trim edges, use a self-piloting bit or separate pilot tip to guide the cut.

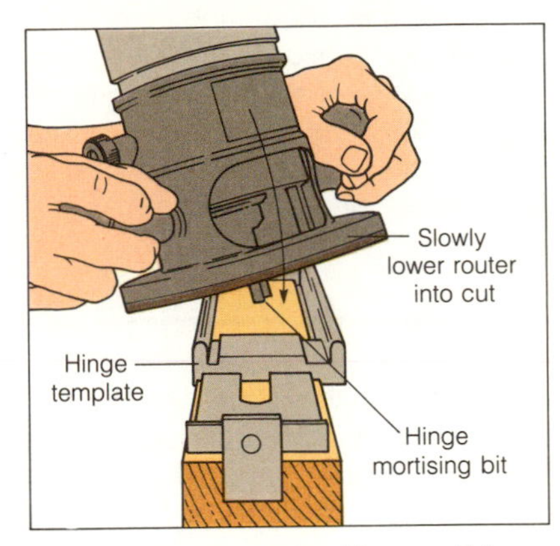

Mortise cuts are made with a mortising or straight bit and guided with a template; plunge cut slowly with the motor running.

Files & Rasps

Generally, files abrade both metal and wood; rasps are for wood only. Perforated rasps, fitted with the proper blade, can shape wood, gypsum wallboard, soft metals, or plastics.

Choosing files or rasps. The factors determining a file's or rasp's performance include tooth pattern, tooth coarseness, length, and shape.

Files are either single-cut or double-cut, as shown at right. Choose a double-cut file for quick material removal and a single-cut for more precise work.

Common coarseness ratings, in descending order from rough to smooth, are *bastard, second cut,* and *smooth cut.*

Rasps differ from files in having individual, triangular teeth. Because they provide more space for wood particles to escape, rasps won't clog as easily as files in soft woods. They are sometimes rated with the same system as files, but often you'll see terms like *wood rasp* (coarsest) and *cabinet rasp* (finest).

The length of both files and rasps affects their coarseness. For example, the teeth on a 10-inch file are correspondingly larger than those on an 8-inch file of the same type.

Common shapes include flat, half-round, and round (rattail). The half-round style makes the best general-purpose file or rasp. Another useful tool is the *four-in-hand,* half file and half rasp on both sides of its half-round profile. Inside curves and holes call for a rattail shape.

Perforated rasps. These inexpensive shaping tools work like cheese graters—the open holes allow shavings to escape, preventing clogs. They're especially good on materials like hardboard or particleboard. The smallest ("pocket") size, 5½ inches, handily performs small jobs like trimming and smoothing gypsum wallboard cuts.

A SELECTION OF FILES & RASPS

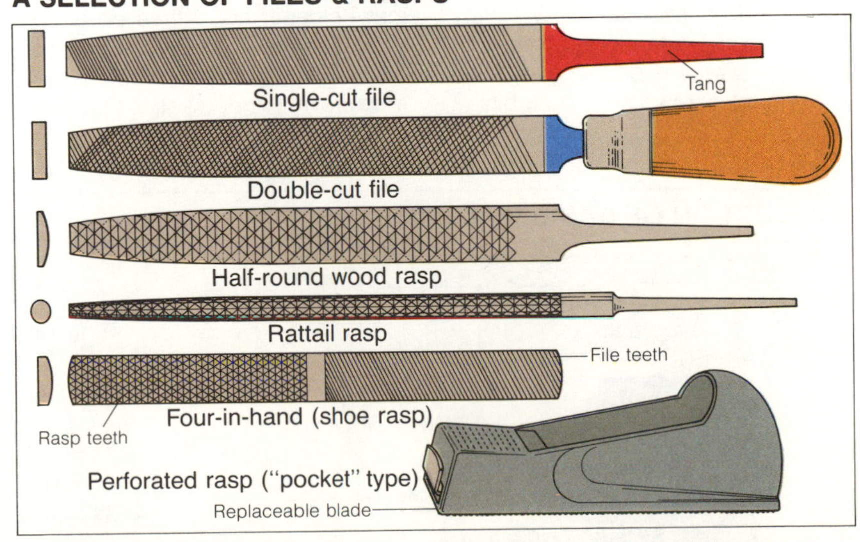

Rasps and files smooth wood; files can handle metal as well. Perforated rasps can be used to shape several materials.

Drilling

Carpenters drill holes for screws, bolts, dowels, doorknobs, locksets and hinges, masonry fasteners, picture hangers—even for nails in hardwoods. And if you double as plumber or electrician, you'll need to drill holes for pipes or electrical cable.

One power tool—the portable electric drill—has virtually replaced its manual counterparts. A reliable model, plus matching bits, may cost little more than a good hand brace, and with the right attachments it's much more versatile.

But many carpenters still find room in their tool boxes for the traditional hand drills and bits. When you're drilling on your back in an awkward basement crawlspace or far from electrical power, you'll begin to understand why.

Portable Electric Drill

An electric drill is classified by the maximum size bit shank accommodated in its chuck (jaws). Three sizes are common: ¼-inch, ⅜-inch, and ½-inch. As chuck size increases, so does power output or *torque*. But the higher the torque, the slower the speed. For most carpentry, the ⅜-inch drill offers the best compromise between power and speed; it also handles a wide range of bits and accessories. If you're drilling large holes in masonry, you'll need the ½-inch drill or a *hammer drill;* both can be rented.

Electric drills are rated light, medium, and heavy-duty (or homeowner, commercial/mechanic, and industrial). For basic carpentry, the medium-duty drill should be fine. If you'll be using it daily or for long, continuous sessions, choose the heavy-duty drill.

Single-speed, two-speed, and variable-speed drills are available. A variable-speed drill allows you to suit the speed to the job—very handy when starting holes, drilling metals, or driving screws. Reversible gears help you remove screws and stuck bits.

Drill bits and accessories. Tool catalogues and hardware stores are jammed with special drill bits, guides, and accessories. Here's a selection of the most reliable and commonly used attachments:

Fractional twist bits, originally designed for drilling metal, are commonly used on wood. Sizes run from 1⁄16 to ⅜ inch (bits over ¼ inch require a ⅜-inch chuck); sets are graduated by 32nds or 64ths. For durability, choose high-speed steel bits. Twist bits also come in *long-shank* "bell hanger" versions, to 18 inches in length, for reaching awkward spots. Or purchase a 12-inch extension bar for your standard bits. *Oversize* twist bits have ¼ or ⅜-inch shanks but bore holes up to ½ inch.

Spade bits drill larger holes in wood—they're typically sized from ⅜ to 1½ inches. When appearance really counts, *brad point* or *power bore* bits are preferred, because they make cleaner holes than either twist or spade bits. They're also more costly. All three bits have center spurs that prevent the "skating" common with twist bits. Brad point bits are available from ¼ inch to an oversize 1 inch; power bore bits range from ⅜ to 1 inch.

For the largest holes, up to 4 inches, the solution is a *hole saw.* Though the type with interchangeable cutting wheels is handy, individual hole saws with fixed blades are more reliable. You'll need a mandrel (arbor)

THE ELECTRIC DRILL

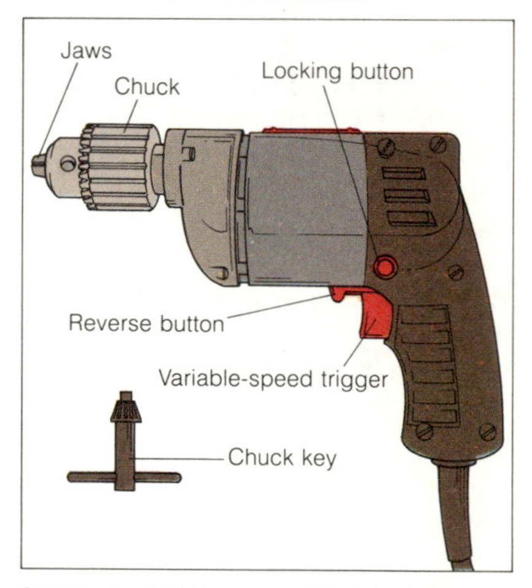

A ⅜-inch drill is a good first choice.

A SELECTION OF BASIC DRILL BITS . . .

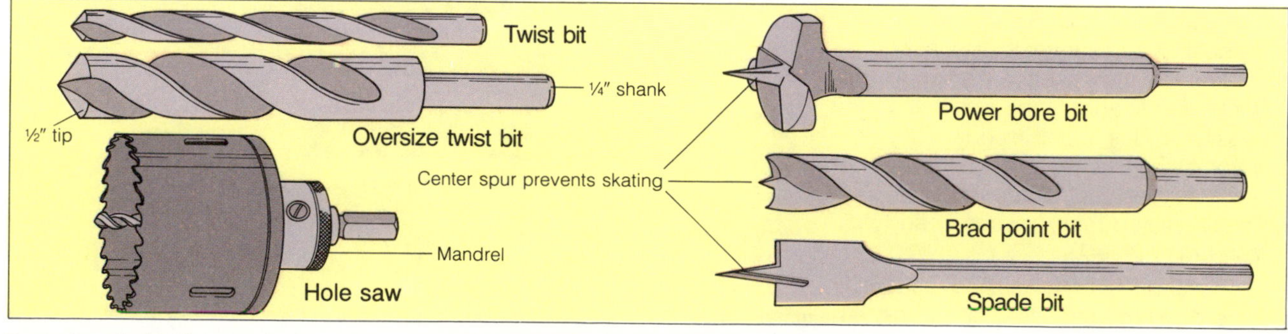

Bore clean holes from 1⁄16 inch to nearly 4 inches in diameter with this basic collection of bits.

DRILLING STRAIGHT HOLES

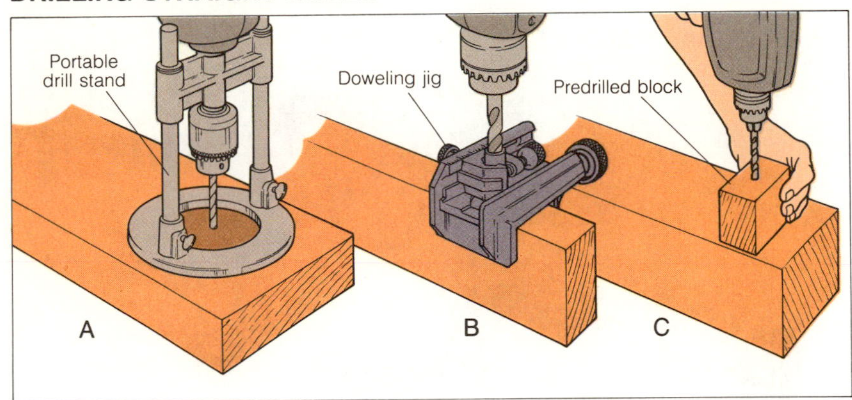

Here are three solutions to the problem of drilling straight, true holes: a portable drill stand can be a handy aid (A); a doweling jig guides holes in a board's face or edge (B); or let a predrilled block guide the way (C).

attachment, as shown on page 18, to fit your drill chuck; the hole saw snaps onto the mandrel. Hole saws up to 2½ inches normally fit a ⅜-inch drill; larger sizes often require a heavy-duty mandrel and a ½-inch chuck. Standard hole saws cut to about ¾ inch, extra-deep models to 2½ inches.

A *pilot* bit drills the proper pilot hole for a screw's threads, a larger hole for its shank, and a countersink for its head—all in one operation. (For more on screws and techniques, see pages 26 and 49.) By drilling deeper with some models, you can form a counterbore hole to help conceal the screw's head. Individual bits prove more reliable than adjustable types. Typical sizes match screws from 3/4 inch by #6 to 2 inches by #12.

Screwdriver and *nutdriver* bits transform your electric drill into a power screwdriver or wrench. A variable-speed drill is a must for these attachments—screws or nuts must be started slowly or they'll strip.

Masonry bits, tipped with tungsten-carbide, chew slowly through concrete, stone, mortar joints, and brick. You'll need a ½-inch or hammer drill to power masonry bits over ⅜ inch.

Drilling tips. When possible, clamp down materials before drilling, particularly when using a ⅜-inch or ½-inch drill. If your drill allows, match the speed to the job: highest speeds for small bits and soft woods, slowest for large bits and metals. As you drill, apply only light pressure, letting the bit do the work. Leave the motor running as you remove the bit from the material. Wear safety goggles, especially when drilling metal.

When drilling large holes in hardwoods or metal—especially with oversize twist bits—first make a smaller lead hole. Back the bit out occasionally to cool it and clear stock from the hole. When drilling through tough metal, lubricate the hole with machine oil as you go.

Special problems. Three main problems plague the drilling process: centering the moving drill bit on its mark, drilling a perpendicular—or correctly angled—hole, and keeping the back surface from breaking away as the drill bit pierces. Here are some time-tested techniques.

Keep a pointed tool to use as a center punch when starting holes. A couple of taps with a hammer and awl or punch will prevent the bit from wandering. A self-centering punch or bit is handy for door hinges.

Several methods and tools can help keep a drill bit going in the path you wish. A portable drill stand, shown above left, is most convenient. Doweling jigs are quite accurate but more costly. Homespun methods include drilling a scrap block and then using the block for a guide. Or simply line up the drill body with the help of a perpendicular try or combination square blade.

To keep the back side of the wood from breaking away, try one of two techniques: lay or clamp a wood scrap firmly against the back of your work piece and drill through the piece into the scrap; or, just before the drill pierces, flip the piece over and finish drilling from the other side.

How do you stop a drill bit at a specified depth? You can buy a stop collar designed for the purpose, use a pilot bit, or wrap electrical or masking tape around the bit at the correct depth.

. . . AND THREE SPECIALTY BITS

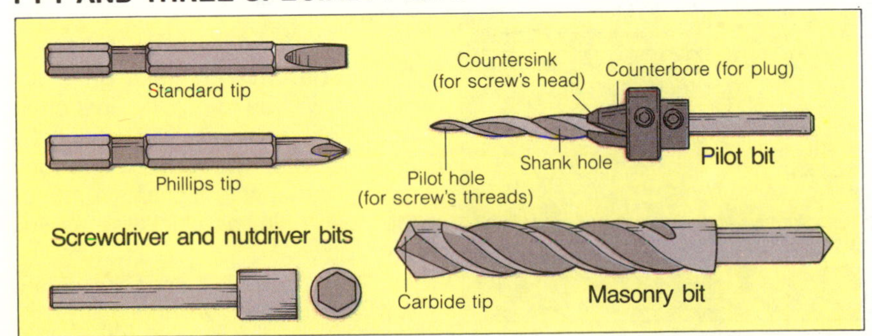

Pilot, screwdriver, nutdriver, and masonry bits allow you to tackle specialized jobs.

. . . Drilling

Hand Drills

If you're reaching for a hand drill, you'll want the hand brace for large holes (to 3 inches in diameter), the eggbeater drill for most screw holes (to ¼ inch), or the push drill for quick holes (to 11⁄64 inch).

Hand brace. Operating the brace is much like turning a crank with an attached bit. Its sweep, which determines the size of the brace, varies from 6 to 14 inches; a 10-inch brace is a good choice. Some braces have a ratchet, a gearlike device that permits you to bore holes in tight places without having to make a full sweep of the brace's body. Many ratcheted braces are reversible.

TRADITIONAL HAND DRILLS

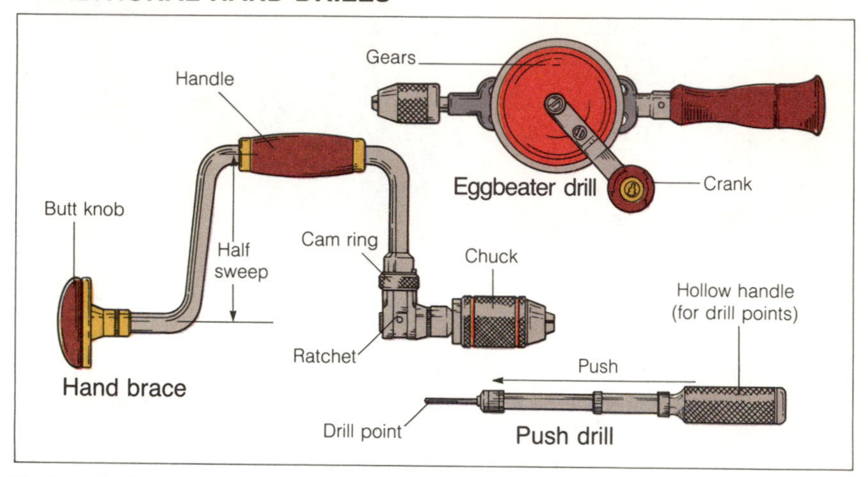

To bore holes up to 3 inches, use a hand brace; eggbeater and push drills bore smaller holes.

■ **Bits for the brace:** Auger bits, standard with a brace, are commonly available in two styles: the double twist or Jennings type, and the single twist/solid center. While the Jennings type bores a cleaner hole, the single twist is both faster and more durable.

Bits also differ in the type and number of cutters and spurs and in the threading on the center screw. For general carpentry, your best bet is a bit with a single twist and solid center, double spurs and cutters, and a single coarse thread.

Standard auger bits range from ¼ to 1½ inches; extra-length bits or extension bars, both up to 18 inches long, help you bore deep holes through thick materials.

Expansive bits, which incorporate an adjustable cutter, drill larger holes from about ⅝ to 3 inches in diameter. Two bits are normally required for this range. Depending on the model, you adjust the cutter either by turning a calibrated screw or by releasing a set screw and manually lining it up. Before trusting the reading, make a test hole in a scrap.

When combined with a screwdriver bit, the hand brace provides extra muscle to help you drive—or remove—large screws. A countersink bit bores a neat taper for a woodscrew's head.

BITS FOR YOUR BRACE

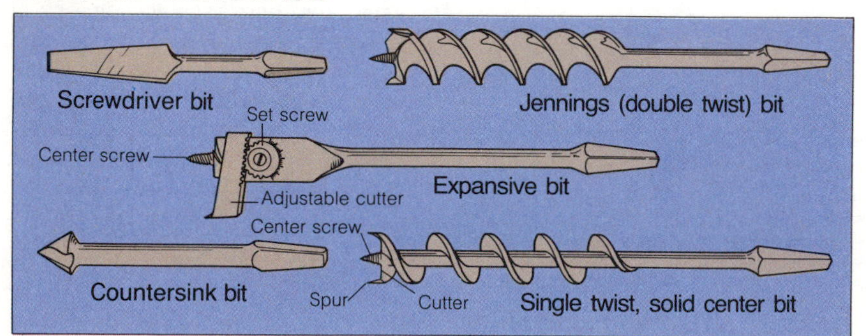

Standard hand drill bits include Jennings and single-twist bits; expansive bits bore holes to 3 inches in diameter.

■ **Tips for using the brace:** To bore a hole, position the bit's center screw on your mark. Holding the round butt knob with one hand, turn the offset handle clockwise with the other. If you're drilling horizontally, use your body to keep the knob in line. It takes some practice to keep the brace from wobbling.

Eggbeater drill. For holes in wood or metal up to 1/4 inch in diameter, the "eggbeater" drill is your tool. You simply aim and crank the handle—hence the name. These drills have a 1/4 or, occasionally, a 3/8-inch chuck capacity. They're often used with standard twist bits (page 18), though some models include a set of drill points (see below) stored in the handle.

Push drill. This compact drill's design allows you to quickly bore small holes with one hand while holding the material with the other. A strong spring-and-spiral mechanism rotates the push drill's bit clockwise as you push down.

The push drill normally comes with a set of two-winged drill points stored in the handle. Sizes range from 1⁄16 to 11⁄64 inch.

Gauging Level & Plumb

One of the carpenter's ongoing concerns is keeping all horizontal surfaces *level* and all vertical surfaces *plumb*. Problems with ill-fitting windows, doors, and finish work can often be traced back to inaccurate leveling and plumbing at an earlier stage. Here's a collection of tools to help keep you in line.

Carpenter's level. Both level and plumb are accurately indicated by a carpenter's level. When an air bubble enclosed in glass tubing at the tool's center lines up exactly between two marks, the surface is level. Similarly, when the instrument is held vertically, tubes near each end indicate plumb.

The standard carpenter's level is 24 inches long with an aluminum, magnesium, or wood body.

Generally speaking, the longer the level, the better: a long body bridges surface contours, producing a more accurate overall reading. *Mason's levels,* up to 78 inches long, are best for checking wall studs and floor joists. As an alternative, you can increase the efficiency of the standard carpenter's level by attaching it to a long, straight board.

Here's how to test a level: place it on a surface that you've determined to be perfectly flat; turn the instrument around and recheck it from the other side. The readings should be identical in each case.

Torpedo level. Typically 9 inches long, a torpedo level can slide into tight spots for local readings—and it slips easily into your tool belt. An even shorter version, the *line level,* hooks onto a taut string that you level and then use for reference when laying out foundation forms, decking, or other large projects.

Water level. Entrapped water always seeks its own level—this principle is put to work in the water level. When plastic tubing is filled with water and the ends are held aloft, the water level at both ends is identical. This helps you to compare the heights of foundation forms and posts more precisely than with a line level, or, on a smaller scale, to transfer heights from one point to another around a room.

Water levels are sold commercially, but you can easily make your own from ¼-inch-diameter clear plastic tubing, food coloring, and water. A pair of rubber stoppers or cork plugs allows you to store the level when not in use.

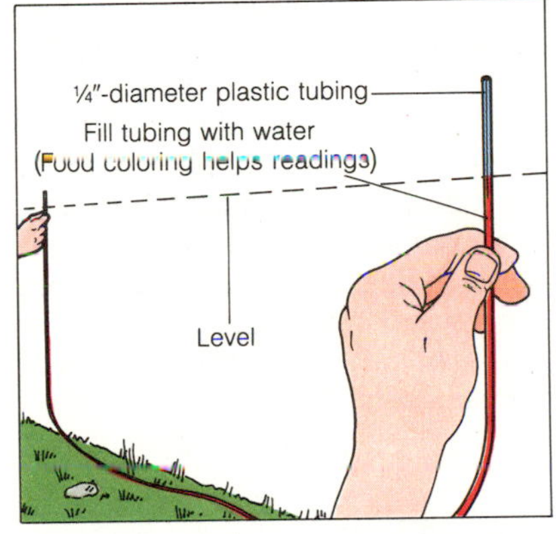

Assemble a simple water level from plastic tubing, water, and some food coloring. When ends are held aloft, water height at both ends is identical.

To operate the water level, first remove the stoppers and make sure that the tubing is free of air bubbles. Have a helper line up the water's height at one end with a reference point—post, foundation wall, or pencil mark. To transfer that point, simply mark the water's height at the other end.

Protractor level. The rotating dial of a protractor level reads off any angle between 0° and 90° in all four quadrants. It's especially useful for matching new work to old. Some carpenter's levels may be fitted with a rotating tube that performs the same function.

Plumb bob. Nothing but a pointed weight at the end of a string, the plumb bob provides a foolproof way to gauge plumb—by gravity. When the other end of the string is fixed and the weight hangs free, the line of the string shows perfect plumb.

The plumb bob is especially good for transferring overhead points to the floor directly below, or vice versa—for example, when you're adding a partition wall or skylight to your present structure. From above, maneuver the weight as close to the floor as possible without touching (it helps to have a partner at the other end), using either your overhead or floor point as a reference. Once the weight stops swinging, it's easy to line up and mark the other point.

BASIC LEVELING TOOLS

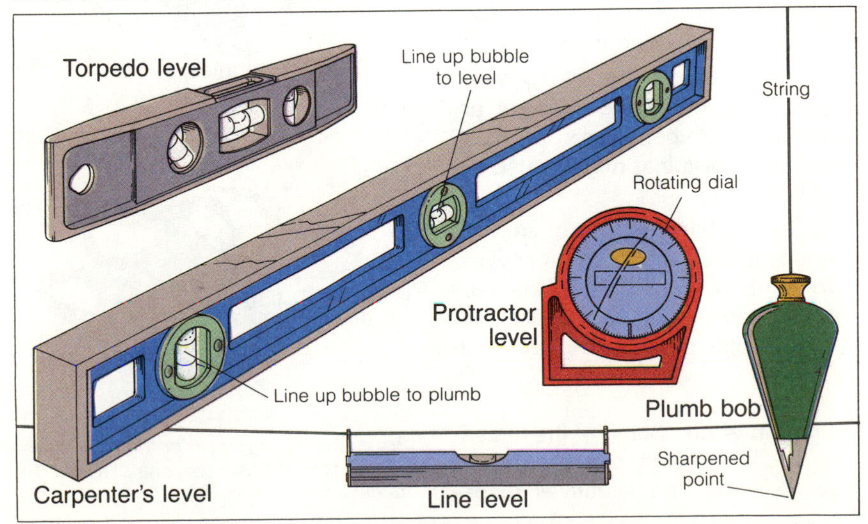

Straight and true—keep horizontal surfaces level and verticals plumb with these tools.

SAFETY WITH TOOLS & MATERIALS

Minimizing or eliminating the risks posed by tools and materials should be standard procedure for every carpenter.

Start your safety campaign by looking at your workplace: Is it clean and organized—or dangerously crowded? Are lighting and ventilation adequate? Next, consider your work clothing and personal safety equipment. Power tools, because of their operating speed, demand special attention, too. And when you're working with electricity, you introduce yet another potential hazard.

Following are some guidelines for safe carpentry.

A safe workplace

A cluttered or poorly lighted workplace invites accidents. And certain materials, if allowed to accumulate, produce potentially toxic fumes, particles, or dust. The following pointers can help promote safety, whether you're on a jobsite or in your home shop.

A clean, well-lighted workplace. Keep tools and materials organized to allow maximum working space. Plan your setup carefully before you begin work.

Whenever possible, avoid working with a partner in cramped quarters. You can too easily be injured by the swing of another's hammer or by a wrecking bar dropped from above.

Carpentry can be messy work. Clean up as you go, preventing an accumulation of bent nails or wood scraps, or spills that might cause uncertain footing.

Good lighting (natural or artificial) leads to neater as well as safer work. Clip-on electric lights, powered by extension cords, make handy supplements; in tight quarters, try a hook-on drop light.

Toxic materials. Some of the materials used in carpentry can pose health hazards: wood preservatives (especially creosote and pentachlorophenol); oil-based enamel, varnish, and lacquer, and the solvents associated with these products; adhesives (especially resorcinol, epoxy, and contact cement); insulation (asbestos fibers and urea formaldehyde); even sawdust or the dust particles from wallboard joint compound.

You can minimize the risks by following a few simple safety rules: 1) Maintain good ventilation to allow particles and fumes to escape from the workplace. 2) Vacuum or wet-mop the area regularly. 3) Read and follow all precautions printed on packaging. 4) Wash skin and workclothes regularly to remove toxic materials. 5) Wear gloves, a painter's mask or respirator, a hat, and safety goggles (all discussed below) to prevent contact with harmful materials.

Safety equipment

Personal safety accessories designed to protect you from injury should be considered basic tools. Here's a complete outfit.

Safety goggles. A must when operating power tools and high-impact hand tools, quality goggles are made of scratch-resistant, shatterproof plastic. Look for a pair that fits comfortably and won't fog unduly—if you can't see clearly through your goggles, they'll impede safety, and if they're uncomfortable, you probably won't wear them.

Work gloves. All-leather or leather-reinforced cotton work gloves are a wise investment. Disposable rubber or plastic gloves are handy for working with solvents, wood preservatives, or adhesives.

Respirator or painter's mask. It's essential to protect yourself from inhaling harmful vapors, dust, or insulation fibers; the heavier the vapors, the better the respirator you'll need. Interchangeable filters are rated for special requirements.

THE BASIC SAFETY OUTFIT

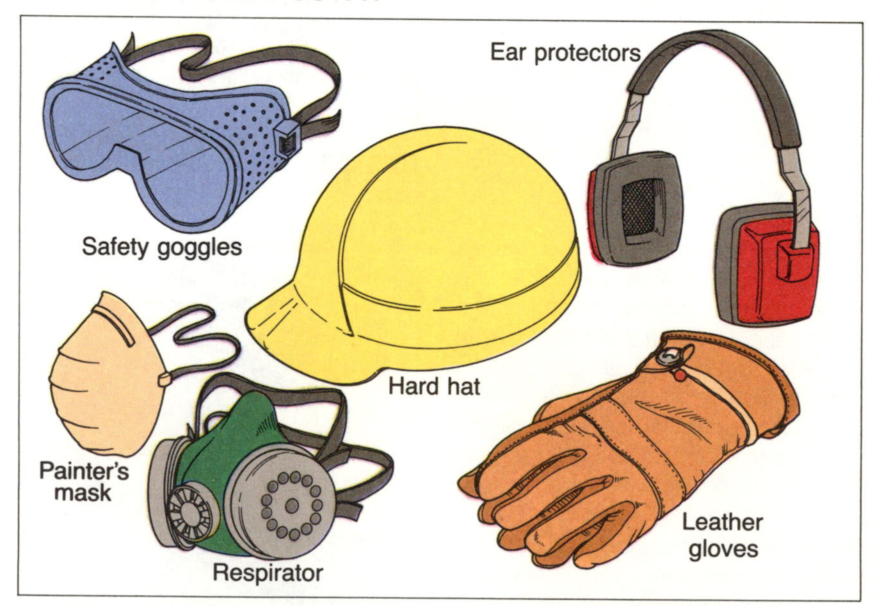

Protective equipment for the eyes, ears, head, hands, and respiratory system should be considered basic carpentry "tools." Sturdy work shoes and clothing complete the safety outfit.

Disposable painter's masks fend off joint compound dust, heavy sawdust, or insulation fibers.

Ear protection. Earplugs or ear protectors are a frequently neglected but crucial piece of equipment. Operating a power tool for any length of time (or even pounding nails in close quarters) will convince you that high noise levels can be painful. They can also cause permanent damage.

The earmuff protectors are most effective; they filter excess noise evenly but still allow you to hear someone speaking.

Protective clothing. Sturdy work shoes, especially steel-toed models, will protect your feet from sharp nails, blades, and dropped tools. (*Always* wear sturdy shoes when framing, especially.)

Any hat will help keep dust and other particles out of your hair, but a hard hat may be a wiser choice—especially when you're working with others in tight quarters.

Safety with power tools

The advent of portable power tools introduced a new potential for injury. If handled with respect, however, these tools are quite safe to operate. The following guidelines will help you establish careful habits.

Checklist for safety. To forestall problems, you'll need to know your tools' capabilities and limitations before you start, and your owner's manuals are the best source of specifics—be sure to read them carefully.

Before you turn on the tool, be sure you're equipped with any necessary supports (see page 10) and clamps for securing the work. Don't wear loose-fitting clothing that could snag in the tool's mechanism; *do* wear safety goggles. If you must use long extension cords in heavily trafficked areas, tape them to the floor.

3 TO 2-PRONG ADAPTER

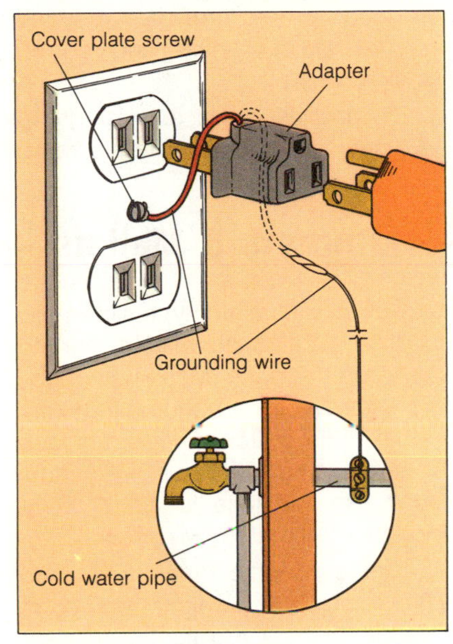

Ground your adapter to the cover plate screw if the outlet itself is grounded. Otherwise, extend the wire to a water pipe (inset).

When you operate a power tool, arrange to do so without interruptions or distractions; keep all visitors, especially young children, away from the work area while the tool is running.

Always keep your power tools sharp, clean, and lubricated according to specifications. And be absolutely certain to unplug any tool before servicing or adjusting it.

Working with electricity. A power tool must be properly grounded unless it's *double-insulated.* Power tools that are neither grounded nor double-insulated can give a serious—even fatal—shock. Never use a single-insulated tool in a damp area or outdoors unless it's properly grounded.

To ground a tool, connect its three-prong plug to a three-hole, grounded outlet. If you have a properly grounded adapter, you can use a two-hole outlet instead, as shown above. Unless the adapter's third wire is grounded itself, you're not protected: if the outlet is properly grounded, simply attach the wire to the outlet's cover plate screw. But if your outlet is not grounded (as is often the case in older homes), you must extend the wire to another grounded object, such as a cold water pipe.

Double-insulated tools are the best defense against a questionable electrical source. These tools contain a built-in second barrier of protective insulation; double-insulated tools are clearly marked and should not be grounded (they'll have two-prong plugs only).

Extension cords. The shorter the extension you can use, the better. A very long cord can overheat and become a fire hazard. And the longer the cord, the less power it will deliver.

The most important factor to consider is the maximum amp load your new extension will need to carry. Every power tool should have a nameplate attached that states its amperage requirement. Add up the requirements of all the tools you plan to plug into the cord at any one time: the extension cord selected *must* have an amp capacity that equals or exceeds this sum. The larger a cord's load capacity, the bigger its wires—and the lower its gauge number.

WHAT CORD GAUGE?

	Extension cord length			
Amperes	25′	50′	75′	100′
2	18	18	18	16
4	18	18	16	14
6	18	16	14	14
8	18	16	14	12
10	16	14	12	12
12	14	14	12	12
14	14	12	10	10
16	12	12	10	10
18	12	12	10	8
20	12	10	8	8

Extension cord's wire gauge depends on the amps and cord length you need. Example: a 50-foot cord for 14 amps requires at least 12-gauge wire.

Fastening

Because the art of fastening is—like measuring and cutting—one of carpentry's fundamentals, you should become well versed in the lore of the tools that make it possible.

Hammers, staplers, and screwdrivers are the most basic. By taking time to master the techniques presented here, you'll reap immediate dividends in the form of a reduced number of bent nails and burred screws.

Wrenches tackle another group of fasteners: bolts and lag screws. Clamps keep the pressure on while glue sets, or hold pieces in place while you nail, drive screws, or drill.

For details on selecting the correct fasteners to complement these tools, see the "Materials" chapter, pages 48–51.

Hammers & Mallets

Everybody's familiar with the claw hammer's appearance, but differences in the tool's shape, weight, and head determine the one you should pick for any specific project.

Carpenters occasionally turn to other tools for driving nails or pounding stubborn joints together. Two you may find handy for this book's projects are the roofer's hatchet and an all-purpose mallet.

Claw hammer. This basic tool is available in two main types: *curved* claw and *ripping* claw. The ripping claw, which is fairly straight, is chiefly designed to pull or rip pieces apart. The curved claw offers more leverage for nail pulling and allows you extra room to swing in tight spots.

Notice that hammer faces may be either flat or slightly convex. The convex, or bell-faced, type allows you to drive a nail flush without marring the wood's surface. Mesh-type faces are available for rough-framing work—the mesh pattern keeps the face from glancing off large nailheads. Don't use this face for finish work, though—the pattern will show.

Head weights range from 7 to 28 ounces. Pick a weight that's comfortable but not too light: your arm may tire sooner swinging a light hammer for heavy work than it would wielding a heavier hammer. For general work, many carpenters choose a 16-ounce curved-claw hammer. For big framing jobs, consider a 20-ounce ripping-claw hammer with a mesh face.

Though both steel and fiberglass handles are stronger, many carpenters still prefer the feel of wood.

■ **Basic nailing techniques:** To start a nail, hold it between your thumb and forefinger below the head and give it a few light taps with the hammer. If you miss the nail, you'll just knock your fingers away rather than smash them.

Once the nail is started, remove your fingers and swing more fully. The most effective hammer stroke combines wrist, arm, and shoulder action. At the top of the stroke, both wrist and elbow are cocked. The downstroke begins with the shoulder and elbow driving down; then, near the stroke's end, the wrist snaps down so that at contact the hammer face, forearm, and upper arm are all roughly parallel to the nailhead. Experiment on some scrap wood until you get the feel—most beginners use too wristy an action.

THREE HAMMERS . . . PLUS A MALLET

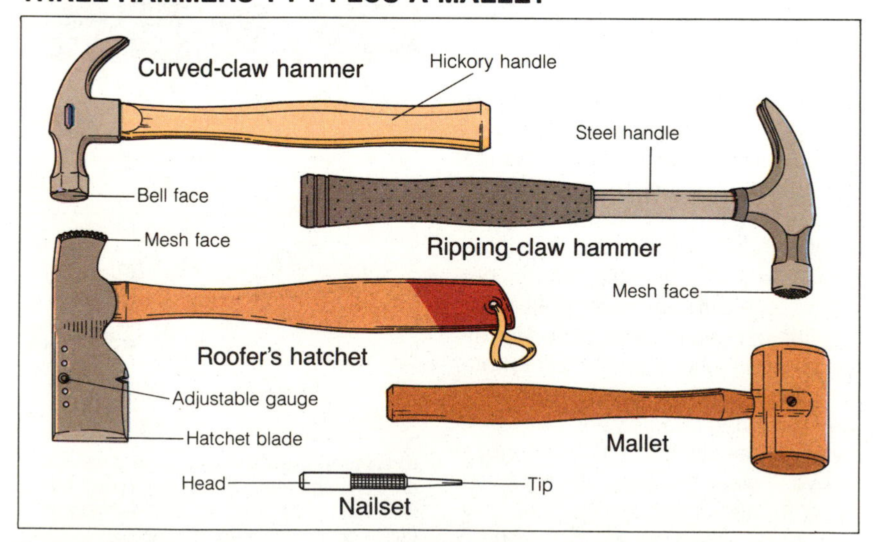

Choose your hammer: a 16-ounce curved claw is the carpenter's favorite; a 20-ounce ripping claw lends authority when driving big nails. Roofer's hatchet and mallet are multipurpose tools.

NAILING TECHNIQUE

A fluid hammer stroke requires that shoulder, elbow, forearm, and wrist work together.

TOENAILING & CLINCHING

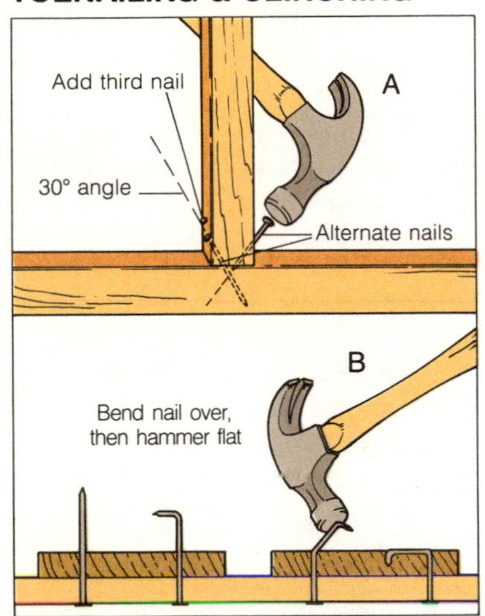

Special nailing situations require you to toenail (A), driving nails at an angle from both sides, or to clinch (B), bending nail ends over, then hammering flat.

PULLING TOUGH NAILS

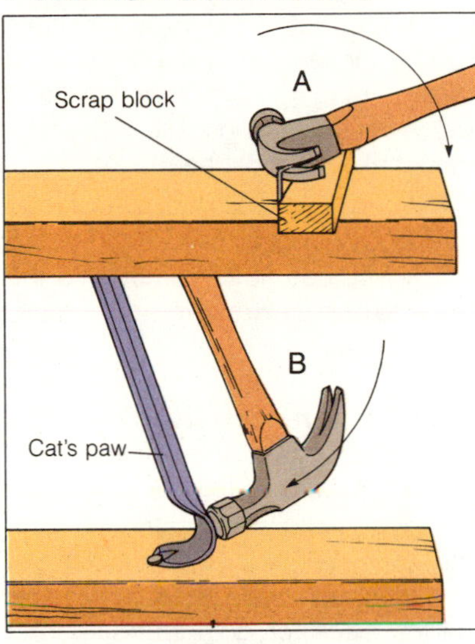

To remove a nail, curl it over a scrap block with a hammer (A), or drive cat's paw (B) or pry bar below the nailhead and pry up.

■ **Toenailing and clinching:** When you can't nail through the face of one board into the end of another (as in nailing wall studs into place), you must nail at an angle through one board into the other. This is called "toenailing." Choosing the proper angle (about 30°) for a strong joint takes practice; the nails should penetrate solidly into both pieces without splitting the wood.

"Clinching" is a method of face-nailing or laminating two boards together. Hammer overlength nails through both boards until their heads are firmly seated. Turn the boards over, bend nail ends across the grain, and hammer them flat.

■ **Driving finishing nails:** Drive finishing or casing nails (see page 48) to within 1/8 inch of the surface, beginning with full hammer strokes and ending with short, careful taps. Then tap the nailhead below the surface with the point of a *nailset*. The resulting hole can be concealed with wood dough or putty.

A set of three nailsets, with tip sizes from 1/32 to 3/32 inch, is handy. Caught without one? Use a stout finishing nail in the same way.

Roofer's hatchet. A multipurpose tool, the roofer's hatchet has both a hammer face for nailing wood shingles and a hatchet blade for scoring shingles and shakes. An adjustable gauge clamps into one of a series of holes to let you quickly determine the proper exposure between shingle courses. (See pages 92–93 for more details.)

Mallets. Wooden, plastic, or rubber-headed mallets are used to knock stubborn joints together, tap dowels into their holes, and drive flooring nailers (see below) or framing chisels without damaging their handles. Mallet weights vary considerably—pick one with enough clout to get the job done, but light enough to maneuver single-handedly.

Staplers & Nailers

Hand and power-driven staplers make some carpentry jobs faster and easier.

Lightweight staplers. These fastening tools are designed to be used one-handed, leaving your other hand free to hold the material. They're especially convenient for fastening building paper, insulation, or ceiling tiles into place. A moderate number of staples, ranging from 1/4 to 1/2 inch or even larger, can be loaded in strips into the stapler's magazine.

To operate the squeeze-type, simply hold the stapler flush against the material and squeeze down on the trigger. With the hammer-type, first position the material where you want it, then strike the stapler against the surface.

Heavy-duty staplers and nailers. Though most heavy-duty models are pneumatically or electrically driven, flooring nailers are driven with a mallet. Some heavy-duty staplers and nailers may hold two or three sizes of fasteners in quantities of 300 or more.

These tools are used primarily for such "assembly-line" jobs as roof sheathing, flooring, or fencing. Because they're expensive, it's wise to rent one for a specific job.

HAND STAPLERS

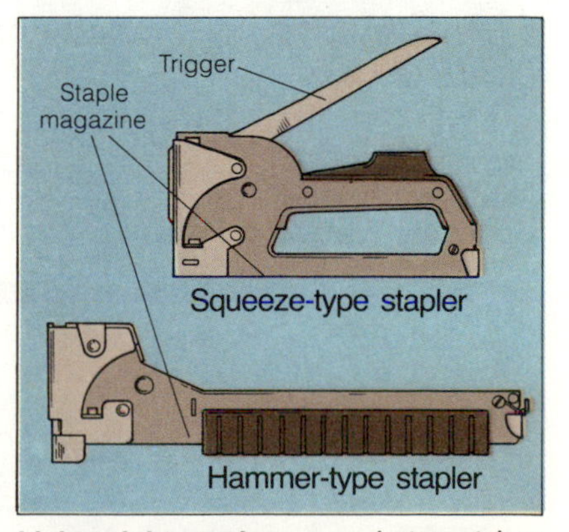

Lightweight staplers come in two styles; either can be operated one-handed.

. . . Fastening

Screwdrivers

The lowly screwdriver vies with the hammer as the most frequently employed tool in a do-it-yourselfer's collection. Yet many prospective carpenters don't know the fine points of selecting and using screwdrivers for first-rate results.

Choosing screwdrivers. Screwdrivers fall into two main categories: standard and Phillips. Within each category, tip width and shank length determine the most efficient driver for the task at hand.

An ill-fitting screwdriver tip leads to a burred screw head or gouged work surface. For fine finish work, look for "cabinet" tips, which have flattened, parallel shapes.

A long screwdriver lets you apply more power than a shorter one. But the long shank may not leave you room to maneuver. Stubby screwdrivers fit into cramped quarters.

Standard screwdrivers have shank lengths from about 3 to 12 inches; corresponding tip widths vary from ⅛ to ⅜ inch. Stock your tool box with three or four sizes covering the range and add a stubby—typically 1½ inches long with a ¼-inch tip.

Phillips screwdrivers, with shanks up to 8 inches, are also sized by tip number, ranging from 0 (the smallest) to 4. A set containing sizes 1, 2, and 3 should answer most needs.

When buying a large, all-purpose screwdriver, choose one with a square shank; you can fit a wrench onto the shank to apply extra leverage to stubborn screws.

Specialty screwdrivers. Two styles of compact *offset* screwdrivers help you reach awkward spots. The standard type, with a 90° bend at each end, comes equipped with either two standard tips (at opposing angles), two Phillips sizes, or both standard and Phillips. The ratcheted type has a tip on each side of its ratcheted head.

Spiral-ratchet screwdrivers accept a variety of tips; they're hand savers when you need to drive quantities of screws. The chuck and tip turn as the handle is pushed down. The larger sizes, though faster, slip easily unless you've had practice.

Screwdriver bits for either electric drill or hand brace (see pages 18–20) also save energy. Electric drills with reverse gears will back screws out, too. (If you choose an electric drill, you'll need a variable-speed model.)

Spring-clip and wedge-tip drivers hold screws while you're starting them; magnetized Phillips drivers serve this function as well.

Driving screws. Screws require predrilled pilot holes in all but the softest materials. Pick a drill bit the diameter of the screw's shank, and drill only as deep as the length of the unthreaded shank. In hardwoods, also drill a smaller hole for the threads below the shank hole; it should be about half as deep as the threaded portion is long. Use a drill bit slightly smaller in diameter than the core between the screw's threads.

Flathead screws (see page 49) are normally countersunk to sit flush with the surface; occasionally, the screw will be sunk even deeper (counterbored) and then covered with wood putty or a plug. An electric drill's pilot bit (see page 19) creates pilot, countersink, and counterbore in one operation.

Try rubbing a bit of soap or wax on the threads of a stubborn screw. If it still sticks, enlarge the pilot hole slightly and try again. To help remove a stubborn screw, try dousing it with a lubricating solvent, or heat the screw's head with the tip of a soldering iron.

A SELECTION OF SCREWDRIVERS

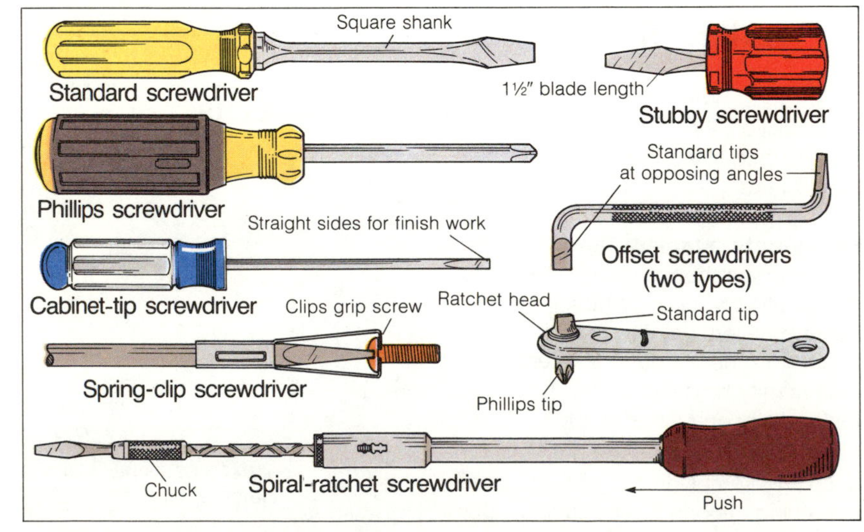

A basic screwdriver collection starts with standard and Phillips types; specialty screwdrivers help reach awkward spots.

PILOT HOLE PROFILE

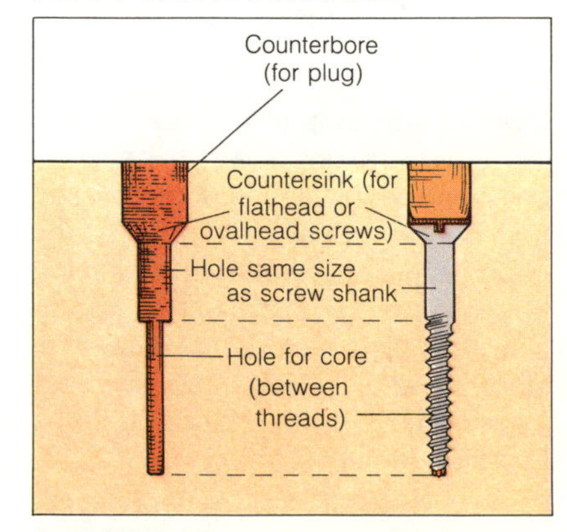

Predrilling saves screws and aching muscles.

Wrenches

You may be surprised how often you'll need a wrench on a carpentry job—to drive lag screws, tighten bolts, or remove existing structures such as cabinets and built-ins.

For occasional use only, choose an *adjustable* wrench, good for a number of bolt or nut sizes. Adjustable wrenches range from 4 to 24 inches in length; more important is the corresponding jaw capacity. For general work, the 10-inch size—with a 1⅛-inch capacity—is a good choice.

Individually sized *box* or *open-end* wrenches are kinder to nuts and bolts. Box ends allow you to apply greatest pressure, but the open end is handy where limited access prevents you from slipping the box end over a bolt. Combination box/open wrenches are also available. Buy standard—not metric—wrenches for carpentry and household use. A typical set ranges from ¼ to 1¼ inches.

A *ratchet and socket* set is the convenient choice for most situations. Three "drive" sizes are common: ¼, ⅜, and ½ inch. The ⅜-inch drive is most versatile; typical socket sizes range from ⅜ to 13/16 inch in increments of 1/16 inch. If you need smaller or larger sizes, adapters will allow you to use ¼ or ½-inch drive sockets. Look for 12-point, rather than 6-point, sockets for general purposes. An extension helps you reach deep spots; a universal joint lets you work at awkward angles.

WRENCHES FOR THE CARPENTER

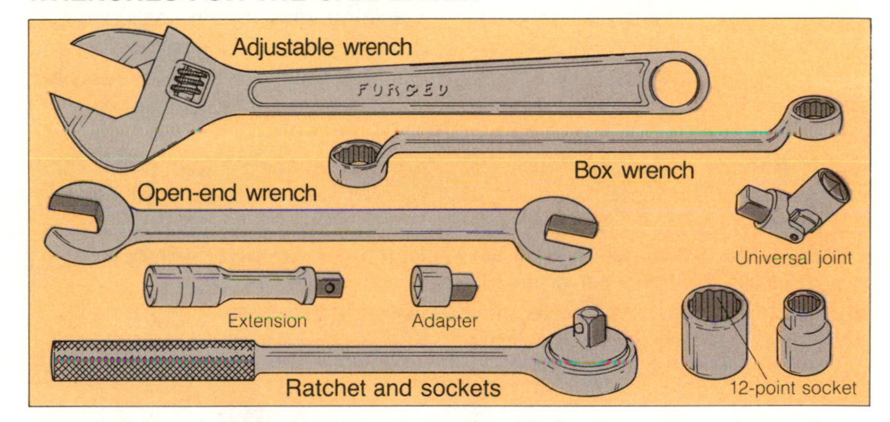

Wrenches drive and remove bolts, nuts, and lag screws.

Clamps

Clamps hold assembled parts tight while glue sets, and practically lend an extra pair of hands when you need them. A carpenter's tip: to protect wood surfaces from being marred by the jaws of a metal clamp, fit a scrap block between the jaws and the wood surface. Tighten clamps until snug, but not too tight.

Clamps come in many shapes and sizes. Here's a selection:

C-clamps are ideal for small jobs—clamping localized areas, holding work to a bench or sawhorse, and attaching scrap guides for cutting. Common jaw widths range from 1 to 8 inches. *Spring* clamps are also handy for fixing scrap guides; in general, they're suitable for quick clamping of light work. Designed like large clothespins, they have jaw capacities of 1 to 3 inches.

The parallel wooden jaws of a *hand screw* adjust for both depth and angle, and easily hold irregular, flat-sided objects. Sizes range from 4 to 16 inches in length, with jaw capacities from 2 to 12 inches.

Bar and pipe clamps have one fixed and one sliding jaw for clamping across wide expanses. Bar clamps are available in lengths up to 6 feet. Typical pipe clamp fittings attach to any length of either ½ or ¾-inch-diameter galvanized or black (nongalvanized) pipe that suits your job; they are much less expensive than their counterparts.

A SELECTION OF CLAMPS

Clamps steady glued pieces until set and hold saw guides while you work.

Finishing

Though it may not be your idea of "hammer and nails" carpentry, most projects require some careful touch-up—prepping, sealing, and sanding—before you're through.

Prepping (rough patching before sanding) is your first step. And exterior seams must be sealed from moisture and air infiltration. When it comes to sanding, you can choose between traditional muscle power and electric power. Hand sanding still produces the finest finish, but power sanders save time and aching muscles.

Here, then, are the tools you'll need before the paintbrush. If you're also the painter, see the *Sunset* books *Wall Coverings* and *Furniture Finishing & Refinishing* for pointers on painting tools, materials, and techniques.

Prepping Tools

Putty knives, taping knives, and caulking guns provide inexpensive aid for neatly and smoothly patching, filling, and sealing.

Putty knives. You can easily fill dents, cracks, and nail holes with a putty knife and wood dough or putty. Blades average 1¼ inches wide and are available in either stiff or flexible styles. Flexing the blade helps drive putty into the hole while leaving a clean, smooth surface.

Joint and taping knives. Best described as stiffer, wider versions of putty knives, joint and taping knives are designed for applying joint compound to wallboard joints and nail dimples (see pages 106–107). Joint knives range from 4 to 6 inches wide, rectangular taping knives to 12 inches. A 5 or 6-inch and a 10-inch knife, along with a corner tool for inside corners, suffice for most jobs.

Caulking gun. Seam-sealing caulking compounds (see page 52) and paneling adhesive are most easily applied with an inexpensive caulking gun.

To operate the gun, first pull the plunger back, handle down, and insert the cartridge. Then push the plunger forward until it engages the cartridge, and turn the handle up. Cut the cartridge nozzle at a 45° angle and pierce the nozzle's seal with a long nail or wire. Work the trigger as needed to control the flow.

TOOLS FOR PATCHING & SEALING

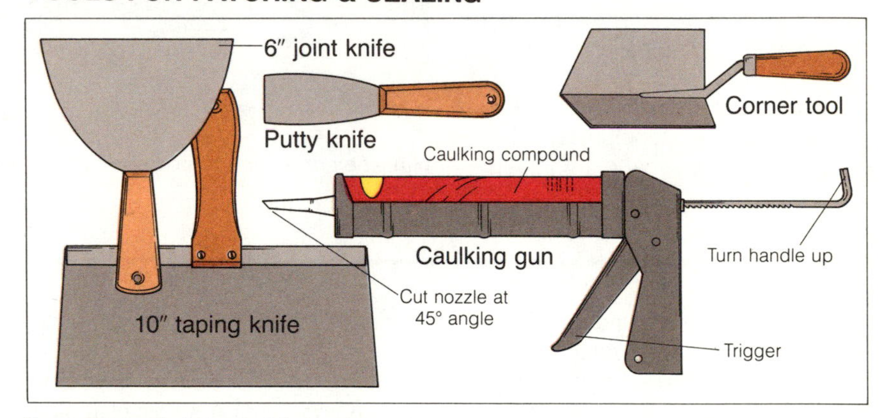

Prepping tools finish nail holes, blemishes, or wallboard joints; a caulking gun fills exterior seams.

Sanding by Hand

Traditional hand sanding still produces the finest surface, and in tight spots or on contoured surfaces, hand sanding may be the only feasible method.

Choosing sandpaper. "Sandpaper is sandpaper," right? Actually, it's not even made with sand. These are the most common materials:

Flint paper, beige colored, offers the least expensive but also least durable option. It's useful for very rough sanding or for removing an old finish.

Aluminum oxide, generally light brown, is a synthetic material of great toughness; choose it for rough to medium hand sanding and for a power sander's belt or pad.

Garnet paper, reddish to golden brown, provides excellent results for hand sanding, especially in the final stages.

Silicon carbide, blue-gray to charcoal, is often termed "wet-or-dry" because its waterproof backing allows you to use it wet, eliminating the clogging tendency of its tiny grains. Try it on metal (wet), as a final "polish" on wood (usually dry), and to sand joint compound when installing gypsum wallboard (wet).

Sandpaper type is usually labeled on the sheet's backing. Other information you'll find there includes grit number, backing weight, and the distinction of open or closed coat.

Grit numbers run from a low of 12 up to 600, but 50 (very coarse) to 220 (very fine) is the common range. Wet-or-dry paper is generally available up to 600-grit.

Backing weights are rated from A (thinnest) to E. In general, backing weight decreases as grit becomes finer.

The terms "open" or "closed" coat refer to the spacing between grit particles. Closed coat paper has more particles to cut faster, but it clogs in soft materials; open coat works better for rough sanding.

Sanding tips. To prepare for a fine finish, divide sanding into at least three stages: rough sand with 50 to 80-grit paper; switch to 120-grit for a second sanding; then sand once more with 180 to 220-grit paper. To provide a flat surface for the sandpaper, buy a sanding block—or make your own from a short 2 by 4 scrap. Always sand in line with the wood grain: cross-grain lines will show up as ugly scratches when finished.

SANDING BY HAND

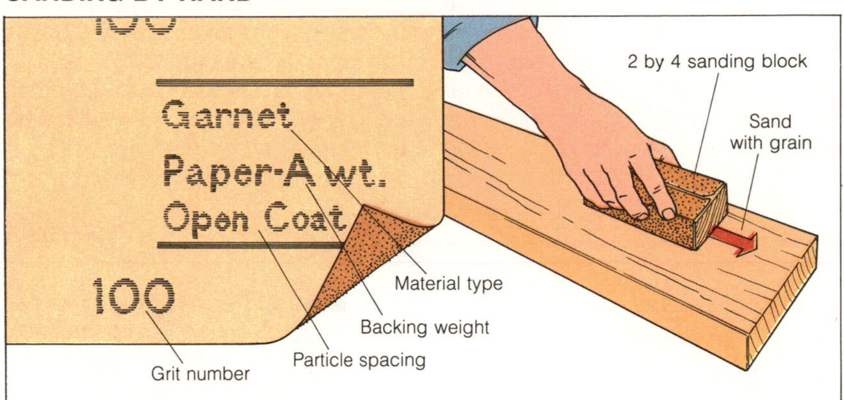

Before selecting sandpaper, check the information on the backing (left). When sanding, use a sanding block (right) in line with the wood grain.

Power Sanders

Portable electric sanders fall into two main types: belt and finishing. Belt sanders abrade wood quickly—they're best for rough, general sanding over large areas. Finishing sanders work at very high speeds to produce a finer, more controlled finish (they won't remove much stock, even with coarse paper attached).

Belt sanders. Sized by the width and length of the belt, the most popular belt sander is probably 3 inches by 24 inches. Features to look for include a dust collection system, as well as convenient methods of replacing the belt and adjusting belt tracking. And be sure you'll be able to find replacement belts for the size you choose.

Belt coarseness was originally rated by fractions from 4¼ (coarsest) to 0 (medium) to 10/0 (finest), but today you'll normally see grit numbers like those on sandpaper sheets. Belt numbers tend to be consistently lower, however: 36 to 50 is coarse, 50 to 80 is medium, and 80 to 120 is fine.

Remember one basic rule for operating a belt sander: always keep the sander moving when it's in contact with the work. Belt sanders can remove a lot of material quickly.

Move the sander forward and back in line with the grain. At the end of each pass, lift it off the surface and start again, overlapping the previous pass by half. For maximum stock removal, you can sand at a slight angle to the grain, but always finish directly with the grain. Don't apply pressure—the weight of the sander alone is sufficient.

Finishing sanders. Although most finishing sanders are available with either straight-line or orbital action, some allow you to switch from one motion to the other. Straight-line action usually produces a finer finish, since the stroke is always back and forth with the grain. An orbital sander moves in very tight circles—up to 12,000 orbits per minute—for a polishing effect.

Finishing sanders range from 4 inches by 4⅜ inches (pad size) to about 4½ inches by 9⅝ inches. The smallest sizes, designed to be held in one hand, allow you to comfortably sand vertical and overhead surfaces. Look for a sander that fits flush into corners or against a wall. Be sure it takes exactly a quarter, a third, or half of a standard sandpaper sheet.

Both straight-line and orbital types work best in line with the grain. For very fine finishes, it's still desirable to do the final sanding by hand.

ELEMENTS OF POWER SANDERS

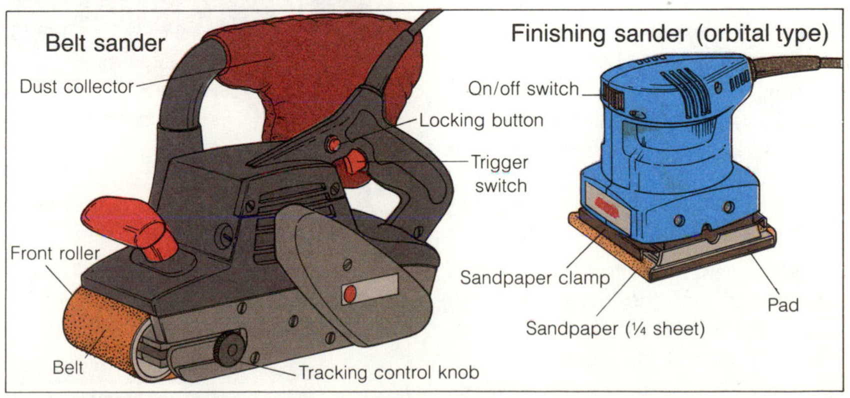

These sanders require little muscle power. The belt sander (left) removes stock quickly; a lightweight finishing sander (right) produces a fine surface.

Dismantling

In a perfect world, carpentry would proceed without a hitch. In reality, the best of us need to pull bent nails or replace a wall stud that's out of plumb. Remodeling calls for another bag of tricks: chiseling through tough plaster and lath, prying up a rotting subfloor, or cutting the opening for a new skylight. And occasionally, dismantling involves some outright bashing to separate old wall coverings from framing members or to knock the framing members themselves apart for removal.

Here, then, are basic tools for carpentry's dirty work. And because the going is often rough and messy, be sure you're equipped with sturdy work gloves, safety goggles, and a painter's mask or respirator (see pages 22–23).

Rough-cutting Tools

These cutting tools are designed to stand up to a variety of tough building materials.

Hacksaw. You'll find that a hacksaw can handle most metal and plastic: nails, pipes, metal lath, and rusty bolt heads. The thin, fine-toothed blades, typically 8, 10, or 12 inches long, can be attached to the hacksaw's steel frame with cutting edge up, down, or horizontal.

If there's no working space for the standard hacksaw, try a "mini" hacksaw (which uses standard blades).

Cold chisels. All-metal versions of the wood chisels shown on page 14, cold chisels accomplish such heavy-duty chopping and splitting chores as shearing rusty bolt heads or breaking through plaster and lath or floorboards.

Standard cold chisels range from 4½ to 8 inches long, with blade widths to 3 inches; long-handled versions are as long as 18 inches. Drive cold chisels with a mallet or single jack (see "Sledges" below).

Reciprocating saw. The number one power tool for roughing-in window, skylight, or door openings, the reciprocating saw operates like a freehand saber saw. Fitted with the proper blade, it cuts such diverse materials as wall studs, plaster and lath, old nails, and steel pipe.

Most reciprocating saws have two speeds; use the lower speed for cutting metal and for finer work, and the high speed for rough cuts in wood. Some models feature variable-speed triggers for more precise control.

Reciprocating saw blades vary in length, tooth size, and design. Generally, blades with 3½ to 10 teeth per inch work best on wood; blades with 14 to 32 tpi are for metal. Other common types include the general-purpose blade (for wood, sheet products, and plaster) and the plaster-cutting blade (which also cuts through the metal lath behind). Blade lengths range from 2½ to 12 inches.

Because the reciprocating saw is an expensive portable power tool, you'll probably prefer to rent one.

When operating the reciprocating saw, keep a firm grip on the rear handle with one hand, and cup the body with your other hand, as shown above right. To prevent excess vibration, keep the blade guard flush against the material being cut. To make a wall cutout, start from a predrilled hole.

Grip the reciprocating saw firmly and keep the blade guard flush against the material being cut.

Saw blades are flexible; this helps you make tricky cuts, but unless you're careful it also leads to snapped blades. It may take some practice to keep the blade tip from bouncing against the work. Be sure to wear safety goggles.

CUTTING TOOLS FOR TOUGH MATERIALS

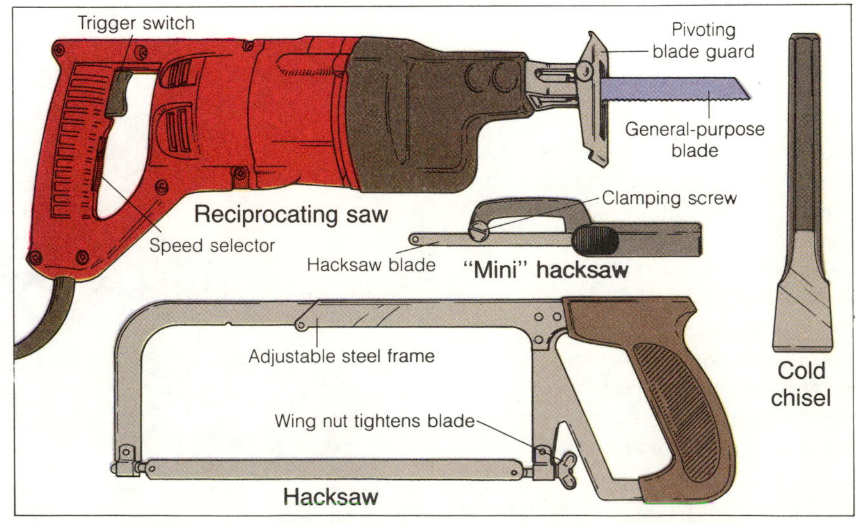

Remove old nails, pipes, and framing with the aid of these rough-cutting tools.

Dismantling Tools

Bending, prying, and some controlled bashing are all facets of the dismantling process. The following tools help supply the muscle.

Pry bars. These bars are handy for removing wallboard or plaster and lath from wall studs, or for pulling exposed nails.

Pry bars come in two basic shapes: flat or hexagonal shank. The hexagonal type, often termed *ripping* bars, may have either an offset end or a straight chisel, as shown at right. The flat type is better for nail-pulling and small cracks, but it tends to flex too much for heavy prying. Typical pry bar lengths are 12 inches for the flat type and 18 inches for ripping bars.

Be sure the pry bar you choose provides a good hammering surface. Drive the chisel end into a crack or under a tough nailhead with hammer blows on the opposite end; pry by hand or with hammer blows.

Wrecking bars. For tough work, you'll need the longer wrecking bar. The advantages of its extended shank are twofold: you'll get much greater leverage, and you'll keep your hands farther from sharp materials like metal lath and nails. Look for a bar with an offset chisel at one end and a hooked, nail-pulling claw at the other end.

You can hammer on the hooked end for extra leverage. Placing a block of scrap wood between the hook and a hard surface not only increases leverage, but protects that surface if you're trying to save it.

Wrecking bars run from 12 to 36 inches; buy a medium size first as a compromise between power and maneuverability.

Sledges. Sometimes it's simplest just to knock one framing member or material away from another. Sledges are oversize hammers for these oversize jobs.

Sized by head weight, sledges range from 2½ to 20 pounds; hardwood handles run from 16 to 36 inches long. Smaller sledges are often termed *hand-drilling hammers* or *single jacks*.

BASIC DISMANTLING TOOLS

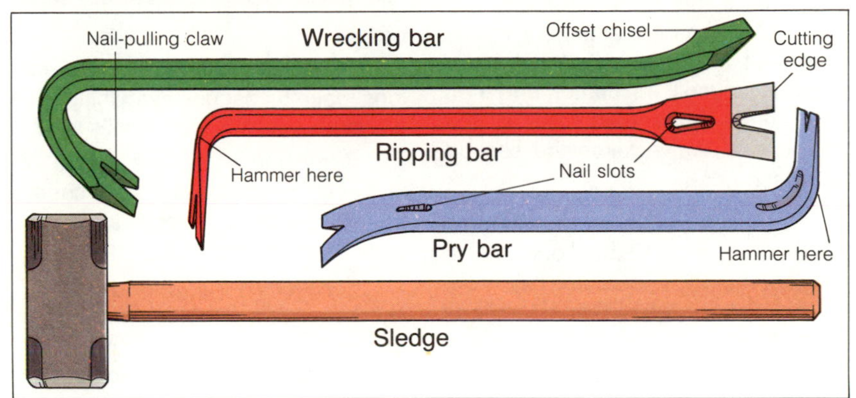

For heavy-duty prying and dismantling, reach for a pry bar, wrecking bar, or sledge.

Pliers. Though pliers aren't strictly prying or dismantling tools, you'll find them invaluable for small-scale gripping, twisting, bending, and cutting. The carpenter can concentrate on three types: lineman's, slip-joint, and locking.

Lineman's pliers perform heavy-duty gripping and bending; side cutters behind the serrated jaws cut wire and thin metals. A typical size is 9 inches.

Slip-joint (or combination) pliers are lightweight versions of lineman's pliers; their distinction is the notch and pivot screw that adjust the jaw capacity. Sizes typically range from 4 to 10 inches; buy a larger size for maximum strength.

Locking pliers keep pressure on the material being gripped, letting you concentrate on turning or bending. They're great for removing gnarled bolts or for holding a nut tight while you turn the bolt's head with a wrench. Typical sizes are 5, 7, and 10 inches, and they come in both curved-jaw and straight-jaw styles.

PLIERS FOR THE CARPENTER

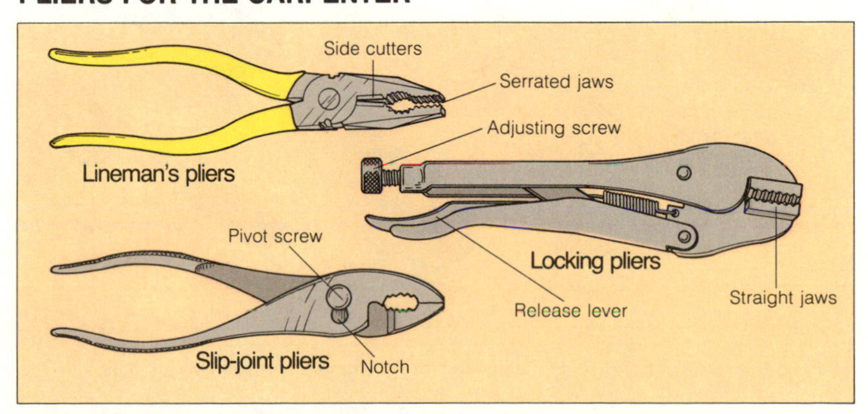

Pliers lend extra muscle for gripping, twisting, bending, and cutting.

THE BASIC—AND MORE ADVANCED—TOOL BOX

What tools do you need to get started? You may find yourself wavering between two strategies: buy nothing until you need it, but then buy every tool required for a particular project; or buy a core of basic tools that will cover most tasks you'll encounter, and then add more sophisticated tools as you need them. We've taken the second approach here.

The items we've chosen tend to be the carpenter's traditional hand tools; power tools and specialty tools for finer finish work make up the "advanced" group. If you're likely to need an expensive power tool only once, why not rent it? If you find yourself needing it again, that may be the time to buy your own.

The page numbers noted in parentheses will lead you to more information on the individual tools.

The basic tool box

The following collection of tools allows you to complete almost any basic project:

Measuring tape, 20 to 25 feet long (page 5), gauges everything from a board's thickness to a wall's length.

Combination square (page 6) lays out cutting lines and checks board square and joint alignment; some models gauge level.

Carpenter's square (page 7) checks square surfaces or corners and lays out long lines, roof rafters, stair stringers, and wall bracing.

Chalkline (page 7) marks long lines more accurately than is possible with a pencil and square.

Crosscut saw (page 8), 8 or 10 point, can handle almost any cutting chore.

Three-quarter-inch butt chisel (page 14), with its short, thick blade, plastic handle, and steel cap, will stand up under a hammer's blows for rough notching and on-the-job trimming.

Block plane (page 15) can be operated one-handed; the small size fits in your nail pouch or tool belt.

Four-in-hand (page 17), with file and rasp teeth on both sides of a half-round profile, gives you four files for the price of one.

Perforated rasp (page 17), available in a pocket model, smooths wallboard cuts and, with the correct blade, trims a variety of materials.

Electric drill (page 18) has virtually replaced the hand drill. A variable-speed, reversible ⅜-inch drill equipped with twist and spade bit sets will handle most tasks.

A CARPENTER'S BASIC TOOL BOX

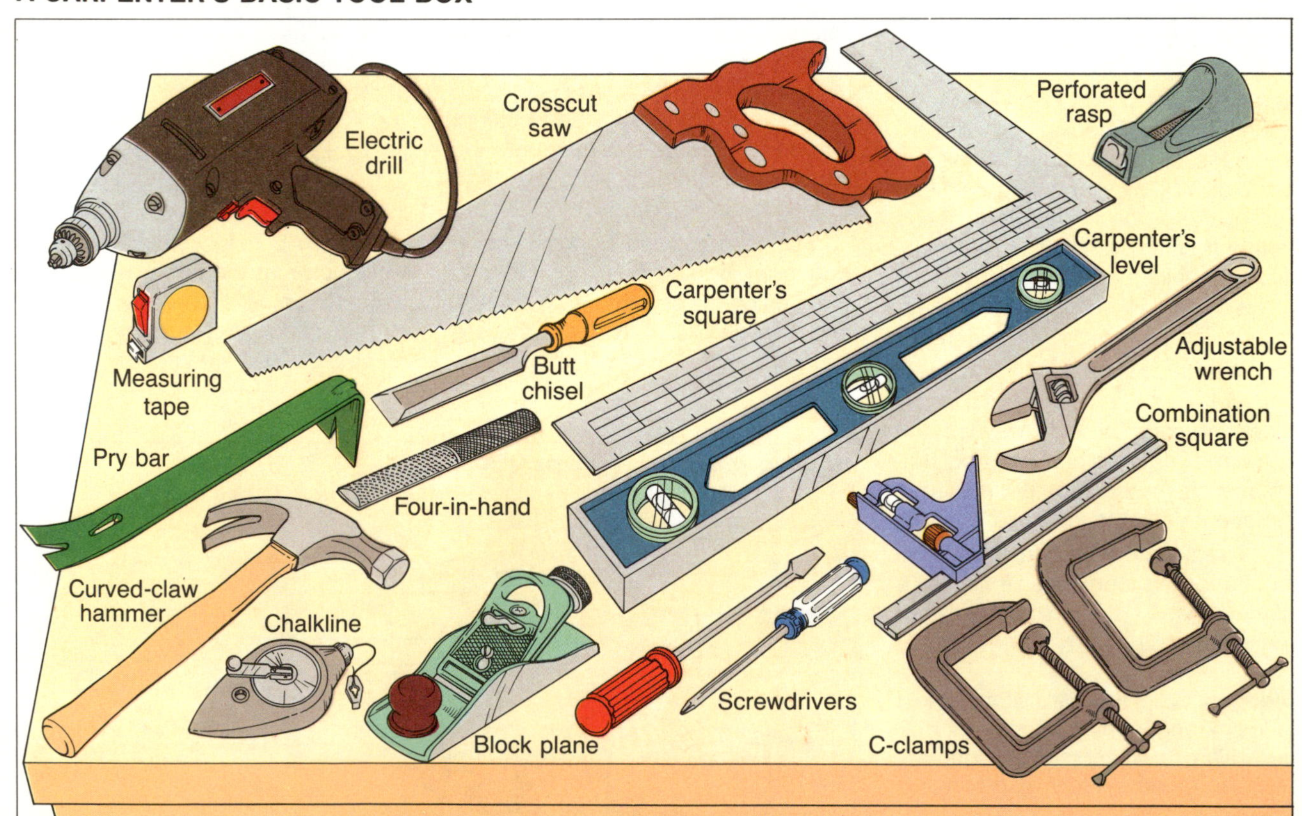

Start with a core of basic tools; this collection will handle most simple projects.

Carpenter's level (page 21), 24 inches long, keeps your work both level and plumb.

Curved-claw hammer (page 24) drives and pulls out nails. A 16-ounce, forged steel head with smooth bell face is a good first choice.

Screwdriver set (page 26) should include two or three standard screwdrivers and a #1 Phillips driver.

Adjustable wrench (page 27) tightens and removes lag screws or bolts. The 10-inch wrench covers a range from 0 to 1⅛ inches.

C-clamps (page 27) hold workpieces together or to a sawhorse and secure saw guides. A set of two 4 to 5-inch clamps is the minimum.

Pry bar (page 31) handles dismantling chores; choose a bar with nail-pulling claw and hammering surface.

The advanced tool box

For special tasks or fine finish work, or simply for faster results, consider these tools:

Adjustable T-bevel (page 6) transfers angles from old work to new—in remodeling projects, for example.

Backsaw and miter box (pages 8 and 10) make fine crosscuts and miter cuts in thin stock—especially molding and trim. A 12 to 16-point backsaw meets most needs.

7¼-inch circular saw (page 11) equipped with combination, rip, and plywood blades makes almost any cutting job faster and easier.

Saber saw (page 13) specializes in cutouts, curves, and scrollwork. Buy a selection of blades for rough-cutting, tight curves, and metals.

Jack plane (page 15), a medium-size bench plane, smooths and squares lumber or doors.

Electric router (page 16) cuts fast, neat grooves or recesses (mortises) and trims edges. Buy bits and guides as you need them.

Plumb bob (page 21) indicates plumb by gravity; it's most useful for transferring points from above or below.

Framing hammer (page 24) helps you do heavy framing work. Consider a 20-ounce ripping-claw hammer with mesh face.

Ratchet and socket set (page 27) drives bolts and lag screws. A ⅜-inch-drive ratchet is your best bet.

Finishing sander (page 29) replaces sanding by hand. Small orbital sanders can be operated one-handed; larger sanders are faster.

ADVANCED TOOLS: ADD AS YOU NEED THEM

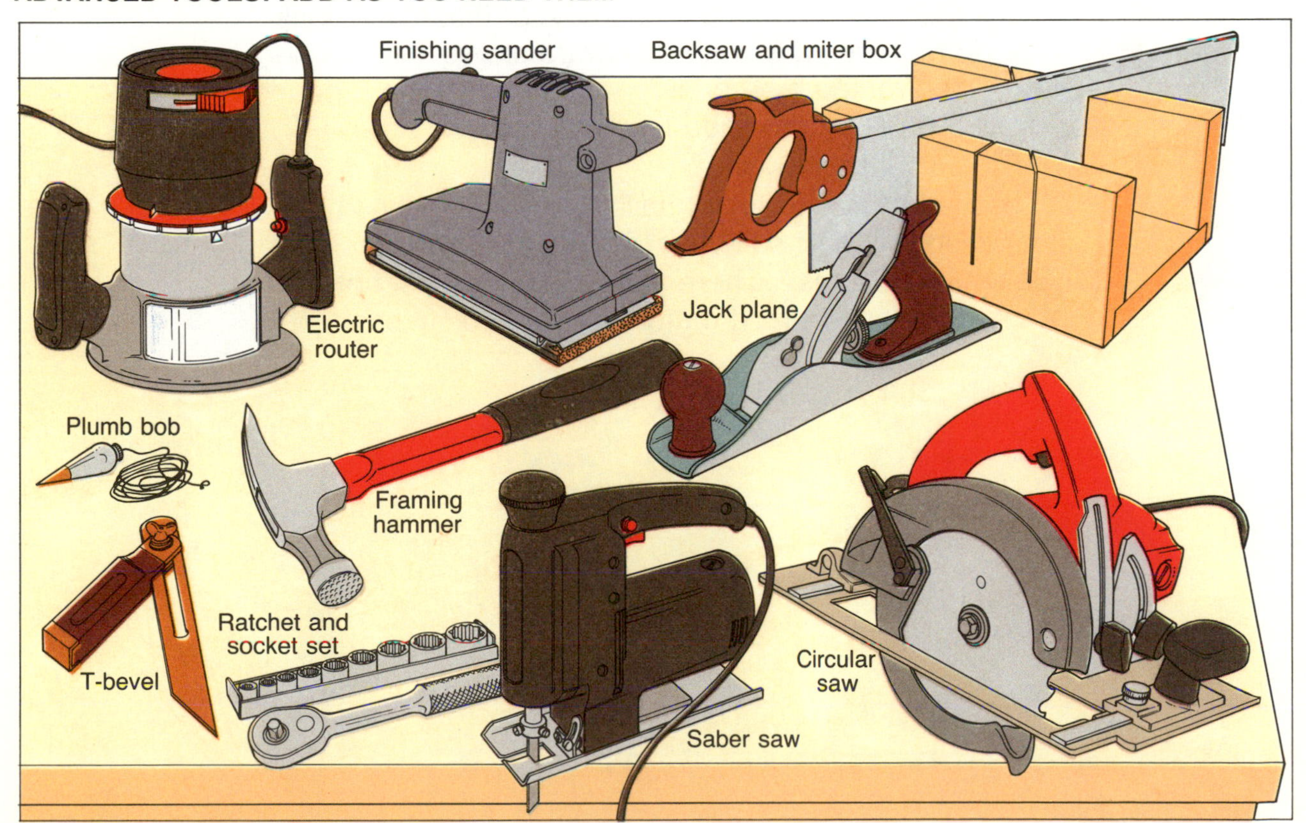

These power and specialty tools make clean cutting and detail work simpler and faster.

Materials

Carpentry's basic materials may seem to lack glamour, but they certainly merit respect. Every carpenter embarking on a project should remember that the finished product will be only as handsome, functional, and durable as the components.

This chapter presents those "everyday" materials of carpentry: lumber, sheet products, paneling and trim, roofing and siding, fasteners, and sealing products. Use this information as a quick primer and reference to what's available and as a dictionary for that mysterious jargon of lumberyards and hardware stores. Then, when you're confronted by a choice—"Do you want a 5-pound box of 3-penny galvanized finishing nails, or don't you?"—you'll be ready.

Just what size finishing nails, or plywood sheathing, or floor joists will best suit your job? Turn to the last chapter, "Basic Procedures," where you'll also find descriptions of the more specialized materials required by each job.

Choosing Lumber

A clean, fresh, uncut board says "carpentry" like no other material. And indeed, lumber is the starting point for almost all carpentry projects. But the landslide of lumber sizes, species, and grades awaiting the uninitiated can be overwhelming at first. You may also be surprised at how "crusty" a busy lumberyard employee can be if you have no idea what you're looking for! On the other hand, armed with an understanding of some basic terms, you can usually secure friendly help with the fine points.

Of course, if you're just building some simple garage shelves or patching a hole in the backyard fence, you can probably close your eyes and buy a couple of boards. But for anything larger, you should do a little homework first.

Lumberyard Lingo

For starters, you'll need to know how lumber is categorized by type and size. Here are some passwords:

Hardwood or softwood? Lumber is divided into hardwoods and softwoods, terms that refer to the origin of the wood: hardwoods come from deciduous trees, softwoods from conifers. The terms can be misleading. Though hardwoods are usually harder than softwoods, some softwoods—like Douglas fir and southern pine—are actually harder than so-called hardwoods such as poplar, aspen, or Philippine mahogany (lauan).

As a rule, softwoods are much less expensive, easier to tool, and more readily available than hardwoods. In fact, nearly all facets of house construction today are done with softwoods. The durable, handsomely grained hardwoods are generally reserved for fine interior paneling, flooring, and other finish work.

Lumber sizing. Lumberyards and lumber grading associations often divide softwood lumber into the five size categories outlined below. In general, softwood lengths run from 6 to 20 feet in 2-foot increments. Hardwoods typically come in standard thicknesses but random lengths: lumberyards sell whatever is available.

Strips are small pieces, less than 1 inch thick and 3 inches wide.

Boards (a standard term for lumber graded by appearance) are normally not more than 2 inches thick, and are 4 to 12 inches wide.

Dimension lumber, graded primarily for strength, is intended for structural framing. These pieces range from 2 to 4 inches thick and are at least 2 inches wide. This category sometimes includes the larger pieces discussed below.

Beams and stringers, structural lumber 5 inches thick or more, have a width at least 2 inches greater than their thickness.

Posts and timbers are heavy construction members 5 inches by 5 inches and larger; width must not exceed thickness by more than 2 inches.

■ **How lumber is sold:** Pieces are sold either by the *lineal foot* or by the *board foot.*

The lineal foot, commonly used for small orders, considers only the length of a piece. For example, you might ask for "twenty 2 by 4s, 8 feet long" or "160 lineal feet of 2 by 4."

The board foot is the most common unit for volume orders; lumberyards often quote prices per 1000 board feet. A piece of wood 1 inch thick by 12 inches wide by 12 inches long equals one board foot. To compute board feet, use this formula: thickness in *inches* × width in *feet* × length in *feet.* For example, a 1 by 6 board 10 feet long would be computed:

1" × ½'(6") × 10' = 5 board feet

And a 4 by 4, 16 feet long, becomes:

4" × ⅓'(4") × 16' = 21⅓ board feet

Of course, you'll still need to list the exact dimensions of the lumber you need so your order can be filled correctly.

Nominal and actual sizes. The beginner's most common stumbling block is assuming that a 2 by 4 is actually 2 inches thick or 4 inches wide. It's not. Such numbers give the *nominal* size of the lumber: its size when sliced from the log. Later, when the piece is dried and surfaced (planed), it's reduced to a smaller size. Almost all softwood lumber you'll find in the lumberyard is surfaced on four sides (designated "S4S"). One notable exception is redwood, which is also sold *rough*—or unsurfaced—for outdoor use. Rough wood remains close to its nominal dimensions, but actual dimensions vary depending on how "green" the piece is (see page 38).

The chart below lists standard surfaced dimensions for common lumber sizes.

STANDARD DIMENSIONS OF SURFACED LUMBER

Nominal size	Surfaced (actual) size
1 by 2	¾" by 1½"
1 by 3	¾" by 2½"
1 by 4	¾" by 3½"
1 by 6	¾" by 5½"
1 by 8	¾" by 7¼"
1 by 10	¾" by 9¼"
1 by 12	¾" by 11¼"
2 by 3	1½" by 2½"
2 by 4	1½" by 3½"
2 by 6	1½" by 5½"
2 by 8	1½" by 7¼"
2 by 10	1½" by 9¼"
2 by 12	1½" by 11¼"
4 by 4	3½" by 3½"
4 by 10	3½" by 9¼"
6 by 8	5½" by 7½"

. . . Choosing Lumber

Lumber Grading Guidelines

Lumber of the same species and size is graded on a sliding scale: the top grade may be virtually flawless; bottom grades are virtually unusable. At the mill, lumber is sorted into grades, then identified with a stamp (see drawing below) or inventoried by species, moisture content, and grade name. (NOTE: Lumberyards sometimes refer to these grades by different names—look for a grading stamp or ask for assistance.)

All grading distinctions are based on defects. One money-saving tip: decide what you can live with and what local building codes require, and buy the lowest acceptable grade.

Softwood grades. You'll find softwoods broken down into two basic categories: strength-graded dimension lumber and appearance-graded boards. The most common scale, employed by the Western Wood Products Association, is shown in the chart below. Here are some added guidelines to help you choose.

■ **Dimension lumber,** rated primarily for strength in house framing, is used for wall studs, sole and top plates, floor and ceiling joists, support beams, and rafters (for a discussion of these terms, see pages 54–61). Fortunately for the carpenter, the American Lumber Standards (ALS) committee has set up guidelines to establish consistent grading categories throughout the country.

Light framing lumber, or the optional *Stud* grade, should be your choice for wall construction in most residential buildings. *Structural joists and planks* is a category that includes joists, rafters, and beams less than 5 inches thick, with a width 5 inches or greater. Unless these components will be exposed in the finished house, you don't need the very top grades—Construction or Select Structural. You must make sure, though, that the grade you choose is accepted by your local building code. Many codes do not allow Utility or Number 3 grades for load-bearing members.

A SAMPLE GRADING STAMP

Look for a grading stamp when you're shopping for lumber. The stamp above, for example, indicates "Standard or Better" dimension lumber with moisture content below 20 percent. (Stamp courtesy Western Wood Products Association.)

THE SOFTWOOD GRADING SYSTEM

Dimension framing lumber (stress-rated)		Boards 1" and thicker (non-stress-rated)		
Light framing 2 by 2 through 4 by 4	Construction Standard Utility	**Appearance grades**	**Selects**	B & Better C Select D Select
Studs 2 by 2 through 4 by 6 10′ and shorter	Stud		**Finish**	Superior Prime E
Structural light framing 2 by 2 through 4 by 4	Select Structural No. 1 No. 2 No. 3	**General purpose boards**		No. 1 Common No. 2 Common No. 3 Common No. 4 Common No. 5 Common
Structural joists & planks 2 by 5 through 4 by 16	Select Structural No. 1 No. 2 No. 3			

Timbers 5" and thicker			
Beams & stringers 5" and thicker Width more than 2" greater than thickness	Select Structural No. 1 No. 2 No. 3	**Posts & timbers 5 by 5 and larger Width not more than 2" greater than thickness**	Select Structural No. 1 No. 2 No. 3

Chart courtesy Western Wood Products Association.

■ **Boards** are graded for appearance; they're not intended for structural framing. If you want a perfect natural finish, buy Select lumber. If you plan to paint, buy a lower grade—paint hides many defects. Common 2 and 3 grades are often chosen specifically for their tight knot patterns. Let your eye be the final judge.

To thicken the plot, certain lumber species, notably redwood and Idaho white pine, have their own grading systems. Look for these grades of redwood, listed in descending order of quality: Clear All Heart, Clear, B grade, Select Heart, Select, Construction Heart, Construction Common, Merchantable Heart, and Merchantable. For Idaho white pine, the categories are Supreme, Choice, and Quality.

Hardwood grades. Like softwood boards, hardwoods are judged by appearance. The number of defects in a given length of hardwood determines its grade. The top grade applies to clear wood at least 8 feet long and 6 inches wide.

The best grades are Firsts, Seconds, and a mix of the two called "FAS." Next comes Select, which permits defects on the back, and Common 1 and 2. Below these, the lumber is generally unusable where appearance counts.

COMPARATIVE GUIDE TO MAJOR SOFTWOODS

Species or species group	Growing range	Characteristics	Major uses
Douglas fir/ western larch	Western states (Rocky Mountains and Pacific Coast ranges)	Very heavy, strong, and stiff; good nail-holding ability. Somewhat difficult to work with hand tools. Number one choice for structural framing.	Framing Sheathing Posts, beams Flooring, subflooring Decks
Southern yellow pine (longleaf, slash, shortleaf, loblolly)	Southeastern U.S. from Maryland to Florida; Atlantic Coast to East Texas	Like Douglas fir, very strong and stiff, hard, good nail-holding ability. Moderately easy to work. High resistance to decay and termites when pressure-treated with preservatives.	Framing Posts, beams Subflooring Interior paneling, trim Decks (if treated)
Eastern white, northern, and western pines (Idaho white, lodgepole, ponderosa, and sugar)	*Eastern and northern pines:* Maine to northern Georgia and across Great Lakes states *Western pines:* western states	Very light and soft woods with below-average resistance to decay and termites, but high resistance to warping. Somewhat weak and limber; moderate nail-holding ability. Shrinkage-prone. Very easy to work.	Framing Exterior siding (if treated) Interior paneling, trim Flooring Cabinetry Shelving
Redwood	Northwestern California, extreme southwestern Oregon	Heartwood known for its durability and resistance to decay, disease, termites. Moderately light with limited structural strength (but strong for its weight). Good workability, but splits easily. Medium nail-holding ability.	Posts, beams Exterior siding, trim Interior paneling, trim Decks Fences Saunas
Western red cedar	Pacific Northwest from southern Alaska to northern California; Washington east to Montana	Similar to redwood in durability and resistance to decay, termites; high resistance to checking, warping, weathering. Strongly aromatic. Somewhat weak and limber; moderate nail-holding ability; very easy to work.	Exposed beams Exterior siding Interior paneling and trim Shingles and shakes Decks Fences Saunas
Hem/fir (eastern and western hemlock; true firs)	*Western hemlock and firs:* Rocky Mountains and Pacific Coast ranges *Eastern hemlock:* Northeastern U.S. and Appalachians	Firs generally lightweight, soft to moderately soft, with average strength. Hemlocks are among least durable softwoods, though fairly strong and stiff; below-average nail-holding ability. Firs are easy to work; hemlocks somewhat more difficult. Shrinkage can be substantial.	Framing Sheathing Flooring, subflooring Interior finish work
Spruce (Engelmann, eastern)	*Engelmann:* Cascades and Rocky Mountains *Eastern:* Maine through Wisconsin, Canada	Lightweight and soft with little resistance to decay. Does resist warping, splitting. Average strength and stiffness; good nail-holding ability; easy to work.	Framing Flooring Interior finish work

...Choosing Lumber

Beyond Grading: How to Pick Lumber

Even within the same stack of lumber, you'll often find striking differences between individual pieces. Whenever possible, sort through the stacks yourself; most lumberyards will let you look and choose if you neatly repile the stacks. Here's what to look for.

Moisture content. When wood is sawn, it's still "green"—that is, unseasoned. Before it's ready for use, most lumber is dried, either by air-drying or kiln-drying.

The highest grades of lumber will normally be stamped "MC-15," indicating a moisture content not exceeding 15 percent. Select and Finish board grades are required to meet this standard. Dimension lumber may be stamped MC-15, "S-DRY" (indicating 19 percent moisture content or less), or "S-GREEN" (meaning green, unseasoned lumber with a moisture content of 20 percent or higher). If you opt for green wood, you're asking for trouble later from splitting, warping, nailpopping, and shrinkage.

Vertical or flat grain? Depending on the cut of the millsaw, lumber will have either parallel grain lines running the length of the piece (vertical grain) or a marbled appearance (flat grain). Vertical grain results from *quarter-sawn* lumber—a cut nearly perpendicular to the annual growth rings (see drawing below). Flat grain results when pieces are *flat-sawn,* or cut tangential to the growth rings. When you can, choose vertical grain lumber; it's less likely to warp or shrink noticeably.

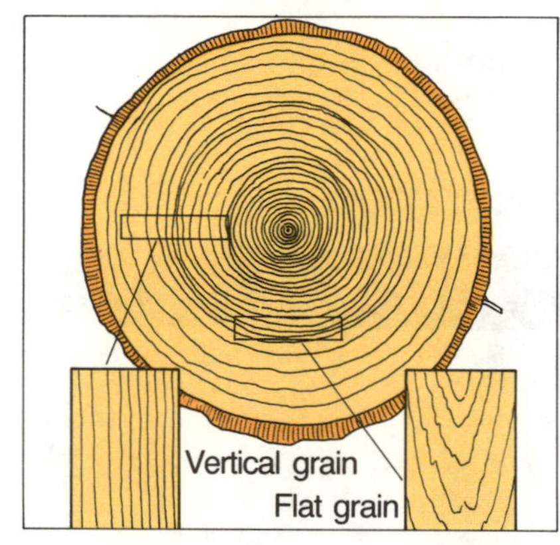

A board's characteristics vary according to how it's cut from the log.

Heartwood or sapwood? The inactive wood nearest the center of a living tree is called *heartwood. Sapwood,* next to the bark, contains the growth cells.

Heartwood is denser and resists decay more efficiently. For this reason, many building codes require that lumber within 6 inches of the ground be heartwood sawn from species naturally resistant to decay—such as redwood, cedar, or cypress. (Or you may often use less durable woods that have been pressure-treated with preservatives—see page 52.)

For most house framing lumber, heartwood and sapwood differences can be ignored. The two usually differ in color, though, and for finish work, you may prefer one over the other.

Weathering and milling defects. Examine the available lumber for defects such as board warpage. Lift each piece by one end and sight down the face and edges. A *crook* (or "crown") is an edge-line warp, a *bow* a face warp. *Cups* are bends across the face; *twists* are multiple bends (see drawings below). Pieces with long, gentle bends can often be made straight when they're nailed.

Other defects to consider include knots, checks, splits, shakes, and wane. Tight *knots* are usually no problem; loose knots may fall out later—if they haven't already. *Checks* are cracks along the annual growth rings in the wood; *splits* are checks that go all the way through the piece; and *shakes* are hollows between growth rings. *Wane* means that the edge or corner of the piece has untrimmed bark or a lack of wood.

Also be on the lookout for general problems such as rotting, staining, insect holes, and pitch pockets (sap reservoirs below the surface).

COMMON LUMBER DEFECTS

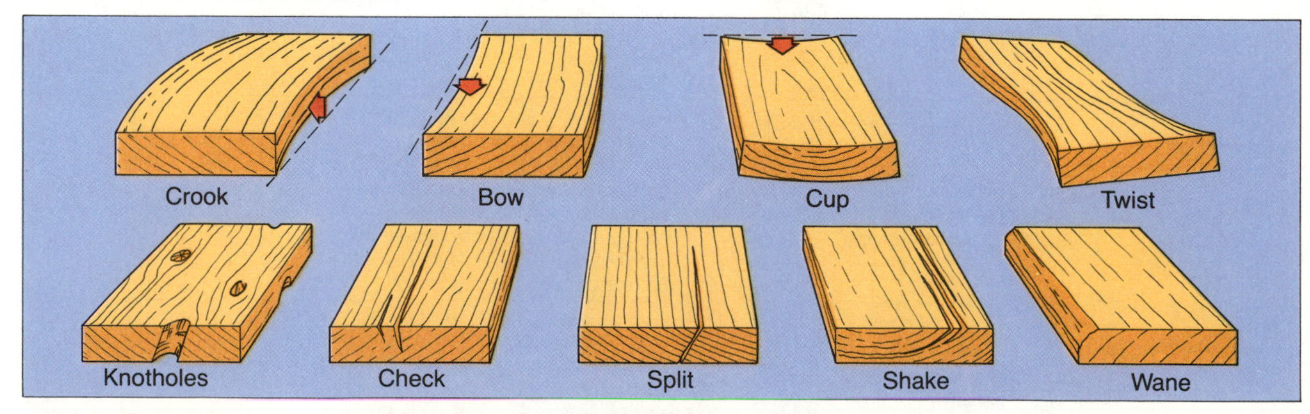

Watch for these flaws at the lumberyard; they aren't usually considered in grading.

HOW TO ESTIMATE MATERIALS

In estimating, the trick is to pinpoint amounts and volumes of materials for your project, always with a slight margin for waste and error, so that you can make all the purchases at once. Done well, estimating saves money. Done carelessly, it will cost both time and money.

Are your plans ready to go? The following guidelines will help you estimate the necessary materials. For tips on shopping and placing your order, see page 63.

Lumber. First of all, you'll want to specify the grade, moisture content, and whether the lumber is to be surfaced or rough. Then figure the *amount* of lumber you'll need. The best way is to simply count up the pieces. A good set of scale plans will come in handy here.

You'll also need to figure the *lengths* of individual pieces. Remember, lumber is normally stocked in even sizes from 6 to 20 feet. Do you need to cover that 10½-foot span with a single 12-foot board, or will 7 and 3½-foot lengths cut from other pieces do just as well? Will the lumber be delivered by truck, or will you have to haul it home on top of your sports car? Carrying a 16-foot piece atop that car is dangerous—perhaps two 8-footers will do?

Plywood. Before figuring bulk coverage of plywood, brush up on the procedures that govern your particular application. (For guidelines on subflooring, wall sheathing, and roof decking, see pages 68, 82, and 88, respectively.)

With that done, sketch the standard 4 by 8-foot panels over the framing system you have planned; then count them. You may need to order plywood by the square foot; if so, multiply the number of panels by 32.

Fasteners. To estimate metal framing connectors, just count the number of posts, beams, joists, or rafters with which you will use the connectors. Once a basic pattern is set, bolts and screws are similarly obvious.

Nails, on the other hand, are difficult to estimate. Probably the best approach is to assess your planned nailing pattern for a given unit, count up the number of nails for that unit, and multiply by the total units—and then add 10 to 15 percent. If you need lots of nails, compare the "per pound" prices of loose versus boxed nails to determine your best buy.

Roofing. Estimates for roofing materials are usually based on the number of *squares* (one square equals 100 square feet) in the roof surface. To find the number of squares in a simple gable roof with unbroken planes, first figure the surface area of each of the roof's rectangles (multiply the length by the width). Next, total up the amounts and divide by 100, rounding off to the next higher figure.

For more complex roof designs, you'll also need to roughly figure the areas of additional rectangles or triangles (see "Siding," below).

Paneling. To estimate solid board paneling, first calculate the square footage to be covered by multiplying height by length. From this figure, subtract the area of all the openings.

Then decide on the pattern of application (see page 109), as well as board width and edge-milling. Remember, a board's actual width is less than its nominal size; you may also need to subtract the width of the edge-milling. Your dealer can help you calculate coverage if you supply the square footage.

If you've chosen sheet paneling and the wall you're covering has a standard 8-foot height, simply measure the length of the wall in feet and divide by the width of your panel—probably 4 feet. Round the figure off to the next higher number. The result is the number of panels you'll need. Unless a very large part of the wall is windows and doors, don't bother trying to deduct materials for them.

Siding. Divide the surfaces to be covered into rectangles (walls) and triangles (gable ends). Round off your measurements to the nearest foot. To determine the area of a rectangle, multiply height by length. To figure the area of a gable end triangle, simply multiply one half the length of the base by the height—in other words, multiply one half the *span* by the *rise* (for an explanation of these terms, see page 76).

You might be able to avoid getting out a ladder to measure the height of your present walls. Measure the width of one board or the length of a shingle exposure, and multiply this by the number of boards or courses from top to bottom.

Add together all the areas that you've computed and then subtract the areas of all openings. To this figure, add 10 percent for waste, plus an additional 15 percent if your house has steep triangles at the gables or any other features that will require extensive cutting.

Popular plywood sidings are sold by the sheet; for other treatments, you'll need to convert your figures to square feet, lineal feet, or board feet. Be sure to take into account the overlap in some styles of siding. Your dealer should be able to supply the right formula.

Insulation. To estimate the necessary amount of insulation for each area, first find the square footage. Next, compute the square footage of doors, windows, and other areas to be excluded; subtract this figure from the first.

If studs or joists are spaced on 16-inch centers, multiply the net square footage by 0.9 to allow for the area taken up by these framing members. If studs or joists are on 24-inch centers, multiply by 0.94. The resulting figure is the total square footage of insulation needed for that area. Each bag or bundle you buy should be labeled with its square-foot coverage.

Sheet Products

Man-made plywood, hardboard, and particleboard have enabled lumber mills to make use of marginal or waste lumber and milling by-products. They offer several advantages over solid lumber: strength in all directions, resistance to warping, ease of installation, and economy.

Construction uses for sheet products in their unfinished state include wall and roof sheathing, subflooring, and underlayment for floor coverings. You'll commonly find them, too, in finish carpentry applications such as cabinetry, countertops, closets, and shelving.

Often, sheet products are imprinted or surfaced with decorative grooves or patterns to enhance their use as exterior siding or interior paneling. For more details on these products, see "Interior Paneling & Trim" (pages 42–43) and "Siding Materials" (pages 46–47).

Plywood

Plywood is manufactured from thin wood layers (veneers) peeled from the log with a very sharp cutter, and then glued together. The grain of each veneer runs perpendicular to adjacent veneers, making plywood strong in all directions. Carpenters use plywood extensively, both for finish work (cabinets, countertops, furniture, and shelving) and for a structural "skin" (sheathing and subflooring) over house framing.

The difference between interior and exterior grades of plywoods lies in the type of glue used to make them, and in the quality of inner veneers. (Exterior grades require weatherproof glue and veneers.) Standard plywood size is 4 feet by 8 feet, though you can find—or special-order—sheets as long as 10 feet. Some lumberyards sell half or quarter sheets.

Like solid lumber, plywoods are divided into softwoods and hardwoods, according to their face and back veneers only.

Softwood plywood. Though softwood plywood may be manufactured from up to 70 species, the most common by far is Douglas fir. Species are rated for stiffness and strength and placed in one of five groups, Group 1 being strongest.

The appearance of a panel's face and back determines its grade. Letters *N* and *A* through *D* designate the different grades (see chart at right for a complete breakdown). Top-of-the-line N grade may need to be special-ordered; use it where you want a perfect natural finish. Generally, an A face is suitable for natural finishes, B for stains, and a repaired C face (called "C-Plugged") for paint.

Plywood comes in many face/back grade combinations, though your lumberyard may stock only a few. A/C (exterior) and A/D (interior) panels are economical choices where only one side will be visible. Face and back grades, glue type, and group number should be stamped on the back or edge of each panel.

When appearance is no factor, you can save money by looking for "shop" plywood, defective panels that don't meet grading standards. Marine plywood, an exterior type with high-quality inner veneers, performs best for areas in constant contact with moisture.

The most common thicknesses of standard softwood plywood range from ¼ to ¾ inch in ⅛-inch increments.

Performance-rated plywood. Some softwood panels are rated for strength in such applications as wall or roof sheathing, subflooring, and concrete foundation forms. Panel thicknesses run from 5/16 inch up to 1⅛ inches; the tongue-and-groove edges available in some types eliminate the need for edge support. Three classifications of exposure durability are available: Exterior (for continuous exposure to the elements), Exposure 1, and Exposure 2.

The grading stamps on these panels supply you with some extra information. On panels intended for all-purpose sheathing, you'll find two numbers separated by a slash (see drawing on facing page). The left-hand number indicates the maximum allowable spacing "on-center" (O.C.)

THE SOFTWOOD PLYWOOD GRADING SCALE

Grade	Description
N	Smooth surface "natural finish" veneer. Select, all heartwood, or all sapwood. Free of open defects. Allows not more than 6 repairs, wood only, per 4 × 8 panel, made parallel to grain and well matched for grain and color.
A	Smooth, accepts paint. Not more than 18 neatly made repairs—boat, sled, or router type, and parallel to grain—permitted. May be used for natural finish in less demanding applications.
B	Solid surface. Shims, circular repair plugs, and tight knots to 1 inch across grain permitted. Some minor splits permitted.
C–Plugged	Improved C veneer with splits limited to ⅛-inch width and knotholes and borer holes limited to ¼ × ½ inch. Admits some broken grain. Synthetic repairs permitted.
C	Tight knots to 1½ inch. Knotholes to 1 inch across grain and some to 1½ inch if total width of knots and knotholes is within specified limits. Synthetic or wood repairs. Discoloration and sanding defects that do not impair strength permitted. Limited splits allowed. Stitching permitted.
D	Knots and knotholes to 2½ inch width across grain and ½ inch larger within specified limits. Limited splits are permitted. Stitching permitted. Limited to interior (Exposure 1 or 2) panels.

Chart courtesy American Plywood Association.

SAMPLE PLYWOOD STAMPS

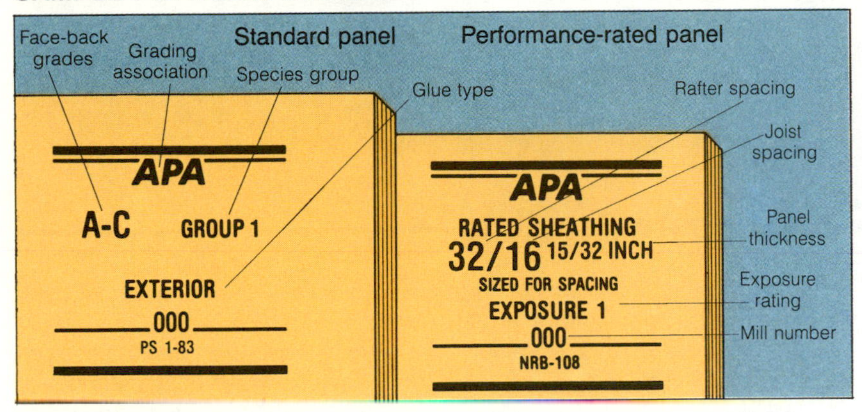

Grading stamps for a standard panel (left) and a performance-rated panel (right) supply all the information you'll need when selecting plywood. (Stamps courtesy American Plywood Association.)

between rafters that support the panel. The second number gives the maximum spacing between floor joists when the panel is used for subflooring. Both are given in inches.

Panels designed for subflooring only are printed with one number, such as "24 O.C."

Be sure to clear all these ratings with your local building officials.

Hardwood plywood. While more expensive than softwood types, hardwood plywoods nonetheless provide an economical alternative to solid hardwoods. Hardwood plywood is identified by the veneer used on the face panel. Popular domestic faces include ash, birch, black walnut, cherry, maple, and oak. A number of imported woods are also available.

Hardwood plywood grading has its own terms; grade, like species designation, refers to face veneer only. Premium grade, the top of the line, has well-matched veneers and uniform color, making it the best choice for a natural finish. Good grade (sometimes designated Number 1) allows less-well-matched veneers and pinhole knots, and normally looks best when stained. Sound grade (Number 2), which still allows no open defects, is best painted. Grades lower than Sound are generally not worth using.

Hardboard

Hardboard is produced by reducing waste wood chips to fibers and then bonding the fibers back together under pressure with natural and synthetic adhesives.

Harder, denser, and cheaper than plywood, hardboard is commonly manufactured in 4 by 8-foot sheets. It may be smooth on both sides or have a meshlike texture on the back. There are two main types, standard and tempered; the latter is designed for strength and moisture resistance. You'll usually see hardboard only in ⅛ and ¼-inch thicknesses. A similar but less dense product, *fiberboard,* is available in thicker sheets but is relatively difficult to find.

In carpentry, the main uses of standard, unfinished hardboard include floor underlayment, cabinet backs and sliding doors, and drawer bottoms. Perforated hardboard, or pegboard, is often combined with metal pegs, hooks, brackets, and racks for hanging storage.

Though relatively easy to cut and shape, hardboard dulls standard tools rapidly. If you plan to work much with hardboard or fiberboard, arm yourself with carbide-tipped saw blades or router bits. Hardboard doesn't hold fasteners well; it's usually necessary to drive them through it into solid wood.

Particleboard

Manufactured from chips and particles of waste wood, particleboard has a speckled appearance, in contrast to the smooth look of hardboard. Standard sheet size is 4 feet by 8 feet; common thicknesses range from ¼ to ¾ inch. Typical uses include floor underlayment, cabinets, and core stock for plastic laminate countertops.

Several types of particleboard are on the market. Most common is a single-layer, matted sheet with uniform density and particle size. But choose the triple-layer type with a denser, smoother face and back if available. An exterior type, called *waferboard,* is designed for wall or roof sheathing, and even for subflooring. (Be sure to check local codes before choosing this option.)

You can work particleboard with standard cutting tools, but equip power tools with carbide-tipped saw blades and router bits. Because the urea or phenol formaldehyde glues used to bond the sheets are potentially toxic, it's prudent to wear a painter's mask while working.

Particleboard won't hold fasteners well. For maximum strength, nail or screw through it into solid wood; otherwise, a combination of nails and glue will give the best results.

Interior Paneling & Trim

The wide spectrum of interior paneling provides many dramatic possibilities for giving your walls just the right look. And there's a product to fit nearly any budget.

You'll discover two main types of wall paneling: solid board and sheet. Board paneling encompasses the many species and millings of both softwoods and hardwoods. The sheet paneling group includes plywood, hardboard, gypsum wallboard, and several other less common materials. You can choose from a wide range of natural, grooved, prefinished, and printed sheets.

Moldings serve two important functions: they provide a "finished," custom look to your paneling, and they hide inaccuracies in joints between materials.

Solid Board Paneling

Solid board paneling is, quite simply, any paneling made up of solid pieces of lumber positioned side-by-side. In some cases, standard, square-edged lumber is used—1 by 4s, 1 by 6s, and so forth. But most of the time, the boards have edges specially milled to overlap or interlock. The three basic millings—square edge, tongue-and-groove, and shiplap—are illustrated below left.

STANDARD EDGE MILLINGS

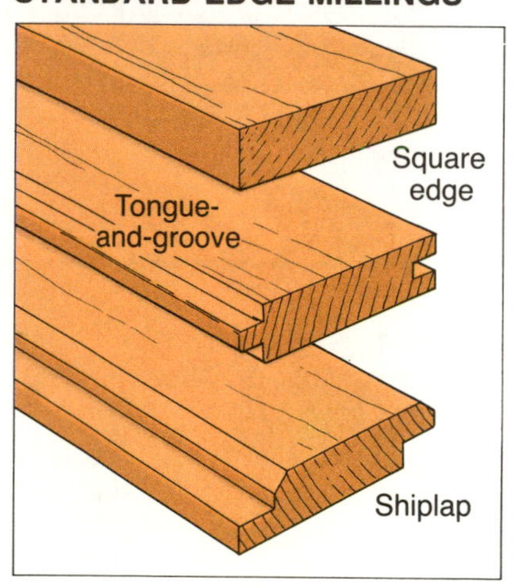

The edges of solid board paneling are typically milled to one of these three shapes.

Thickness of paneling boards ranges from ⅜ to ⅞ inch. The most common thicknesses, though, are ½ inch and ¾ inch (1-inch nominal size). Board widths range from 3 to 12 inches; but remember, these are nominal—not surfaced—sizes. For actual widths of surfaced lumber, see page 35. (Actual paneling widths may vary somewhat, depending on the milling.) Standard board lengths range from 6 to 20 feet.

What are the choices? Hardwood boards are milled from such species as birch, cherry, mahogany, maple, oak, pecan, rosewood, teak, and walnut. Common softwoods include cedar, cypress, fir, hemlock, pine, redwood, and spruce.

For grading purposes, most board paneling can be termed *clear* or *knotty.* Clear softwood paneling boards normally correspond to any Select or Finish board grade (see the chart, page 36), knotty panels to Common 2 and 3 grades. Some species have their own designations. For example, when choosing redwood, look for Clear All Heart and Clear, or (for knots) B grade and Rustic.

No matter what the grade and milling, boards may be surfaced either smooth or "resawn" (rough). Another option is "barnwood," made from softwoods like redwood or cedar. Some barnwood is simulated; another type is actually weathered boards salvaged from unpainted barns and shacks. You can also choose any board type and milling designed for exterior siding.

To prevent shrinkage and cupping, look for kiln-dried boards. And before installing any board paneling, it's ideal to "condition" the boards by stacking them—in the room to be paneled—for 7 to 10 days prior to installation. This allows the paneling to fully adapt to the room temperature and humidity, preventing warping or buckling after installation.

Sheet Paneling

Sheet Paneling is a catchall term for wall paneling that comes in large, man-made panels—most commonly 4 feet by 8 feet.

Plywood paneling. Generally speaking, any of the standard, unfinished plywood sheets described on pages 40–41 may be used for wall paneling. Both softwood grades from N down to C-Plugged and hardwood grades from Premium to Sound offer face veneers without open defects. Redwood grades include Premium and Select (Custom). Specialty-grade hardwood plywood gives you the luxury of custom-matching veneers within the panel or between adjacent panels.

But standard panels aren't your only option. Prefinished and vinyl-faced decorative styles are available, and resin-coated panels are especially designed for painting. Plywood face textures range from highly polished to resawn; many types feature decorative grooves or shiplap edges.

In addition to the standard 4 by 8-foot sheets, you can find some types in 4 by 9 or 4 by 10-foot dimensions. The standard thicknesses of interior sheet paneling are ¼ inch or ⁵⁄₁₆ inch; avoid thinner panels—they're difficult to work with and not very durable. Exterior plywood sidings—ideal choices where you want rustic, rough-sawn textures and patterns—come primarily in two thicknesses: ⅜ inch and ⅝ inch. The ⅜ inch thickness is ample for interiors.

Hardboard paneling. Hardboard makes a tough, pliable, and water-resistant paneling. Sold in 4 by 8-foot sheets, it ranges in thickness from 3⁄16 to 3⁄8 inch; 1⁄4 inch is standard.

The most common surface finishes are imitation wood; generally grooved to look like solid board paneling, wood imitations are available in highly polished, resawn, or coarser brushed textures.

In addition to the wood finishes, you can find panels embossed with a pattern—basket weave, wicker, or louvered, to name a few.

For installation around tubs and showers, you'll find hardboard with a vinyl or plastic-laminated finish that's easily cleaned and sheds water. The plastic-laminated type is most durable. In utility areas, standard hardboard or pegboard (see page 41) makes thrifty, paintable paneling.

Gypsum wallboard. The most common wall and ceiling material, gypsum wallboard serves either as an inexpensive finish paneling or as a backing for other paneling materials. Essentially, it's compressed gypsum dust sandwiched between a smooth, paintable paper face and a heavier paper backing.

Joints between standard wallboard panels are normally concealed with perforated fiber tape and joint compound. Then panels may be painted, wallpapered, or coated with a textured surface treatment. Sound tiring? Factory-decorated, vinyl-faced varieties are available for use as finished panels.

Panels are 4 feet wide; lengths vary from the standard 8 feet up to 14 feet. Common thicknesses are 3⁄8 inch for a backing material for other paneling, 1⁄2 inch for final wall coverings, and 5⁄8 inch to meet fire code requirements (for example, where walls border a garage space). Choose water-resistant wallboard, identified by green or blue paper covers, in the bathroom or kitchen sink area or wherever moisture may collect.

Moldings and Trim

The absence of moldings, it's often said, is a sign of good craftsmanship. Even in the most basic, wallboard-paneled room, though, moldings have their place along the base of walls and around door and window frames. Many traditional architectural styles make extensive use of moldings; if you're matching styles or simply trying to introduce some flair, moldings may be the answer.

Traditional wood moldings come in many standard patterns and sizes. You can buy them natural, prefinished (painted or stained), or vinyl-wrapped with decorative printing. Common lengths range from 3 to 20 feet. Natural or stained moldings are normally continuous, clear lengths. Painted and vinyl-wrapped moldings are often shorter lengths finger-jointed at their ends to make longer pieces. Ponderosa pine is the most popular molding stock.

Besides wood moldings, you can choose from wood-grain-printed plastic, vinyl, or aluminum. Many dealers carry a wide variety of these moldings and color-matched nails to complement whatever style of paneling you choose.

When ordering moldings, remember that thickness is specified first, width second, and length last (for example: 3⁄8 inch *thick* by 2 1⁄4 inches *wide* by 8 feet *long*). Both thickness and width are measured at their widest point.

The drawing below shows many of the standard molding patterns and their common applications. But remember, you can also choose strips or boards—for example, clear redwood 1 by 4s or 1 by 6s—for handsome, bold trim over either solid board or sheet paneling.

POPULAR MOLDING PATTERNS

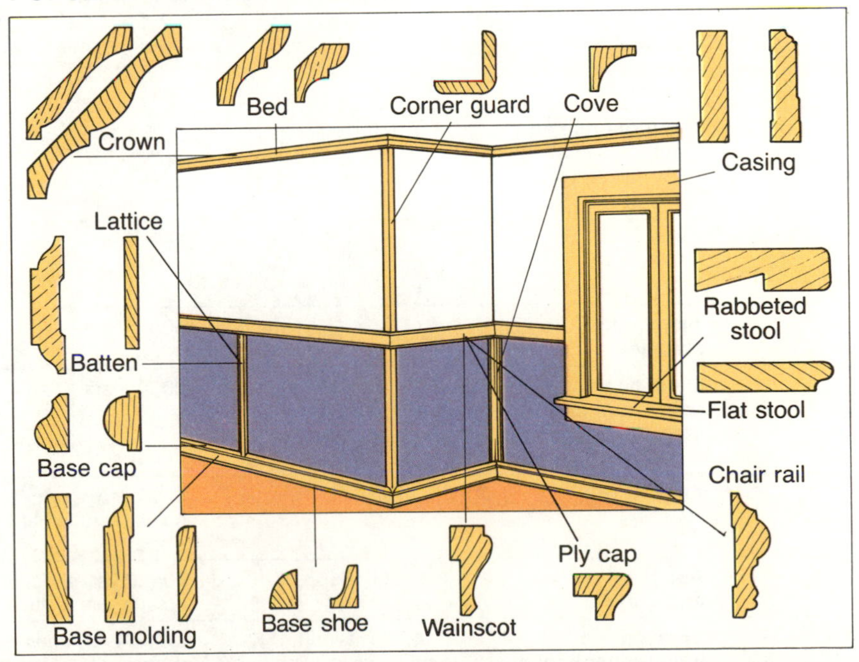

Moldings hide uneven joints between wall, ceiling, and floor coverings while creating a decorative appearance. The most common types and their uses are shown above.

Roofing Materials

Wood, asphalt, and tile—the traditional choices for rooftop protection—today find competition from aluminum, galvanized steel, slate, and even plastic or sod. Flat or very low-sloping roofs may have surfaces of either tar and gravel, or a polyurethane foam (contractor-applied only) that provides insulation as well as weather-proofing.

The chart below provides a direct comparison of the roofing materials most commonly available. You'll find materials to meet nearly every budget and architectural style. Here are some additional terms and factors to consider, whether you're just browsing through the chart or shopping in earnest.

Hidden roofing components. Roofing materials are not fastened directly to rafters. So when thinking "roofs," you'll also need to consider decking, underlayment, and flashing. Decking—the nailing base that's attached to roof rafters—may be either solid (plywood sheathing or tongue-and-groove boards) or open (usually spaced 1 by 4s). Underlayment, in the form of asphalt-impregnated roofing felt, lies atop the decking. Flashings, most commonly made from a malleable, 28-gauge sheet metal, provide extra protection at vulnerable roof spots like valleys, chimneys, drip edges, and vents. For details on all these components, see pages 88–89.

Fire ratings. Underwriters' Laboratory has tested all types of roofing materials for such qualities as resistance to igniting, support for the spread of fire, and emission of burning brands (that add to a fire hazard).

Materials are rated Class A (those with the most fire-retardant qualities), Class B, and Class C. Because of their flammable nature, materials such as untreated wood shingles or shakes receive no rating.

Roof slope. The slope, or pitch, of a roof refers to the vertical rise measured against a standard horizontal distance of 12 inches. For example, the term "4 in 12," applied to a roof, tells you that the roof rises 4 inches

BUYER'S GUIDE TO ROOFING MATERIALS

Material	Weight per square (Pounds)	Durability	Fire rating
Asphalt shingles (felt base)	240–300	12–20 years, depending on sun's intensity	C
Asphalt shingles (fiberglass base)	220–300	15–25 years, depending on sun's intensity	A
Wood shingles and shakes	144–350	15–25 years, depending on slope, heat, humidity	None, untreated; C, when treated with fire retardant; B, with use of fire retardant and foil underlayment
Tile (concrete and clay)	900–1000	50+ years	A
Slate	900–1000	50+ years	A
Aluminum shingles	50	50+ years	C or better
Metal panels (aluminum or steel)	45–75	20+ years	C or better
Asphalt roll roofing	90–180	5–15 years, depending on water runoff at low slope	C or A
Asphalt ("tar") and gravel	250–650	10–20 years, depending on sun's intensity	C
Sprayed polyurethane foam	20 for 1-inch thickness	Life of building with proper maintenance	A

vertically for every 12 horizontal inches. As a rule, do-it-yourselfers should not attempt to work on roofs with slopes steeper than 6 in 12.

To measure a roof's slope, you'll need a carpenter's level and a tape measure or ruler. Mark the level 12 inches in from one end. Resting that end on the roof, raise or lower the opposite end to obtain a level reading. The distance between the roof and the bottom of the level at the 12-inch mark is your slope.

"Square." Roofing materials are often sold by the square, a roofer's term indicating the amount of roofing material (allowing for overlap) needed to cover 100 square feet of roof.

Shopping tips. The slope of your roof and local building codes will quickly eliminate the materials that aren't suitable for your project. Compare the remaining materials in terms of the following qualities: *cost versus durability* (premium-quality asphalt shingles, for example, may be less expensive than standard asphalt shingles on a cost-per-year basis); *warranties* (some provide only material replacement, others also cover labor); *ease of application*; *availability* (shipping or delivery charges may greatly inflate the price of some materials); and *appearance*.

When you order. To save yourself the backbreaking labor of loading materials onto the roof, have them delivered there directly if at all possible—especially if you're putting down shingles or shakes over open sheathing (see pages 88–89). Most roofing suppliers have hydraulically operated scissors trucks specially designed for rooftop deliveries.

Try to arrange to have materials delivered just before it's time to install them. If materials must be kept on the site for an extended period, it's best to store them indoors in a dry place, away from extreme temperatures. Materials left outdoors should be stacked off the ground on 2 by 4s and covered with plastic to protect them from the elements.

Recommended minimum slope	Merits	Drawbacks
4 in 12 and up; down to 2 in 12 with additional underlayment	Available in wide range of colors, textures, Standard and Premium weights; easy to apply and repair; conforms easily to curves in roof surface; low maintenance; economical.	Less durable and less fire-resistant, though equal in cost to fiberglass base shingles.
4 in 12 and up; down to 2 in 12 with additional underlayment	Durable and highly fire-resistant; available in wide range of colors, textures, Standard and Premium weights; easy to apply and repair; conforms to curved surfaces; low maintenance.	Brittle when applied in temperatures below 50°F/10°C.
4 in 12 and up; down to 3 in 12 with additional underlayment	Appealing natural appearance with strong shadow lines; durable. Choose #1 ("Blue Label") shingles for roofing. For more on both shingle and shake types, see pages 92–93.	Flammable unless treated with fire retardants; treated wood expensive; time-consuming application.
4 in 12 and up; down to 3 in 12 with additional underlayment	Extremely durable; fireproof; available in flat, curved, and ribbed shapes; moderate range of colors.	Costly to ship; difficult to install; requires sufficient framing to support weight; cracks easily when walked on.
4 in 12 and up	Attractive, traditional appearance; impervious to deterioration; fireproof; available in several colors.	Expensive and costly to ship; difficult to install; requires sufficient framing to support weight; may become brittle with age.
4 in 12 and up	Lightweight; fire-resistant; made to resemble wood shakes; moderate range of colors.	Can be scratched or dented by heavy hail, falling tree branches.
1 in 12 and up	Aluminum: lightweight; durable; maintenance-free for prepainted panels; sheds snow easily. Steel: strong; durable; fire-resistant; sheds snow easily.	Contraction and expansion of metal can cause leaks at nail holes; noisy in rain.
1 in 12 and up	Economical, easy to apply.	Drab appearance.
0 in 12 and up	Most waterproof of all roofing materials.	Must be professionally applied; difficult to locate leaks; black surfaces absorb heat.
0 in 12 and up	Continuous membrane produces watertight surface; good insulation value; lightweight; durable when protective coating is maintained.	Must be professionally applied; quality depends on skill of applicator; deteriorates under sunlight if not properly coated.

Siding Materials

Where the protection of your house is concerned, siding is every bit as important as a good roof. But siding offers more than protection. The color, texture, and pattern create the "look" your house presents to the world, so you'll want to pick a material that wears well from an esthetic, as well as a practical, point of view.

The range of wood-based siding materials includes traditional solid boards, exterior plywood, hardboard, and shakes and shingles. Add sidings of aluminum, steel, and vinyl, and the choices can make you dizzy. The chart below offers a closer look at all these materials. (Other favorites—brick, imitation stone and brick, and stucco—aren't discussed, being beyond the scope of basic carpentry.)

Shopping for siding. If you're in the market for new siding, remember that the relative cost of a particular siding

A COMPARISON OF DO-IT-YOURSELF SIDINGS

Material	Types and characteristics	Durability
Solid boards	Available in many species. Redwood and cedar have natural resistance to decay. Milled in a variety of patterns. Depending on type, may be applied horizontally, vertically, or diagonally. For more information on patterns and application, see chart on page 84. Nominal dimensions are 1" thick, 4" to 12" wide, random lengths to 20'. Bevel patterns are slightly thinner; battens may be narrower. Sold untreated, treated with water repellent, primed, painted, or stained.	30 years to life of building, depending on periodic maintenance.
Exterior plywood	Most typical siding species are Douglas fir, western red cedar, southern pine, and redwood. Face veneer determines designation (interior veneers may be mixed). Broad range of textures, from rough or resawn to smooth overlay for painting. Typical pattern has grooves cut vertically to simulate solid board siding. **Sheets** are 4' wide, 8' to 10' long. They are applied either vertically or horizontally. **Lap boards** are 6" to 12" wide, 16' long. Thicknesses of both: ⅜" to ⅝". Sold untreated, pretreated with water repellent, primed, painted, or stained.	30 years to life of building, depending on maintenance.
Hardboard	Available smooth or in textures including rough-sawn board, stucco, and many more. **Sheets** are 4' wide, 8' to 10' long. They are usually applied vertically. **Lap boards** are 6" to 12" wide, 16' long. Thicknesses of both: ⅜" to ½". Sold primed, primed and painted, or opaque stained.	30 years to life of building, depending on maintenance. Prepainted finish guaranteed to 5 years.
Cedar shingles and shakes	Mostly western red cedar; some eastern white cedar **Shingles** are distinguished by grade: #1 ("Blue Label") are the best; #2 ("Red Label") are second best and quite acceptable as an underlayment when double coursing. Also available in a variety of specialty patterns. **Shakes** are available in four shapes and textures, varied by scoring, sawing, or splitting. Also sold in the form of "sidewall shingles," specialty products that are basically shingles with heavily machine-grooved surfaces. Widths are random, from 3" to 14". Lengths: 16" (shingles only), 18", and 24". Shakes are thicker than shingles, with butts from ⅜" to ¾" thick. For saving labor, shingles and shakes come prebonded on 4' and 8' plywood panels. Primarily sold unpainted. Also available prestained or painted.	20 to 40 years, depending on heat, humidity, and maintenance
Vinyl	Extruded from polyvinyl chloride (PVC) in white and pastel colors. Smooth and wood-grain textures are typical. Horizontal panels simulate single 6" and 8"-wide lap boards. Vertical panels simulate single 8"-wide boards with battens. Other styles: single panels that simulate two 4", 5", or 6"-wide lap boards, or three 4"-wide lap boards. Standard length: 12'6".	40 years to life of building.
Aluminum	Extruded panels in a wide range of factory-baked colors, textures. Types and dimensions are the same as vinyl. Also sold as 12" by 36" or 48" panels of simulated cedar shakes.	40 years to life of building.

type varies widely from region to region. Also, when you're shopping for solid board or exterior plywood, keep an eye out for the same general characteristics and grading scales by which you'd judge standard lumber (pages 35–38) or plywood (pages 40–41).

Sheathing, building paper, flashing. Because many siding products cannot be fastened directly to wall studs, your costs may include sheathing, building paper, and flashing. Sheathing is required under some siding types to increase their rigidity, provide a solid nailing base, and add structural strength. Building paper adds an extra layer of protection over sheathing. Flashings keep water away from door and window frames and seams between plywood or hardboard siding panels. For more details on all these components, see pages 82–87

Maintenance	Installation	Merits and drawbacks
Ends should be treated with water repellent before installation. Needs painting or opaque staining every 4 to 6 years, transparent staining every 3 to 5 years, or finishing with water repellent every 2 years.	Difficulty varies with pattern. Most are manageable with basic carpentry skills and tools.	**Merits:** Natural material. Provides small measure of insulation. Broad range of styles and patterns. Easy to handle and work. Takes a wide range of finishes. **Drawbacks:** Burns. Prone to split, crack, warp, and peel (if painted). Species other than redwood and cedar heartwoods are susceptible to termite damage when in direct contact with soil, and to water rot if not properly finished.
Before using, seal all edges with water repellent, stain sealer, or exterior house paint primer. Restain or repaint every 5 years.	Sheets go up quickly. Manageable with basic carpentry skills and tools.	**Merits:** Easy to apply. Provides small measure of insulation. Can serve as both sheathing and siding, adding great structural support and strength to a wall. Less expensive than wood boards, yet offers same type of appearance. Broad range of styles and patterns. **Drawbacks:** Burns. May "check" (show small surface cracks) or delaminate from excessive moisture. Susceptible to termite damage when in direct contact with soil, and to water rot if not properly finished.
Before using, seal all edges with water repellent, stain sealer, or exterior house paint primer. Paint or stain unprimed and preprimed hardboard within 60 days of installation; then repaint or restain every 5 years.	Sheets go up quickly. Manageable with basic carpentry skills and tools.	**Merits:** Uniform in appearance, without typical defects of wood. Easy to apply. Numerous surface textures and designs. Takes finishes well. **Drawbacks:** Does not have plywood's strength or nail-holding ability. Susceptible to termite damage when in direct contact with soil, and to water rot and buckling if not properly finished. Cannot take transparent finishes.
In hot, humid climates, apply fungicide/mildew retardant every 3 years. In dry climates, preserve resiliency with oil finish every 5 years.	Time-consuming because of small pieces, but manageable with basic skills and tools plus a roofer's hatchet.	**Merits:** Rustic look of real wood. Provides small measure of insulation. Easy to handle and work. Easy to repair. Adapts well to rounded walls and intricate architectural styles. **Drawbacks:** Burns. Prone to rot, splinter, crack, and curl; may be pried loose by wind. Changes color with age unless treated.
None except annual hosing off.	Manageable with basic carpentry skills and tools, plus zipper tool, snap-lock punch, and aviation shears or circular saw.	**Merits:** Won't rot, rust, peel, or blister. Burns, but won't feed flames. Easiest of synthetics to apply and repair. Resists denting Scratches do not show. **Drawbacks:** Only white and pastel colors available. Doesn't take paint well. Sun may cause long-range fading and deterioration. Brittle when cold.
Needs annual hosing off. Clean surface stains with nonabrasive detergent. Refinish with paint recommended by the manufacturer.	Manageable with basic skills and tools, plus aviation shears or circular saw and brake tool for bending trim. Use aluminum nails only.	**Merits:** Won't rot, rust, or blister. Fireproof and impervious to termites. Lightweight and easy to handle. **Drawbacks:** Dents and scratches easily. May corrode near salt water.

Fasteners

Nails, screws, bolts, and adhesives—it is with these that the carpenter assembles all the other materials we've discussed in this chapter.

The fastest way to join two pieces is to nail them together, so nails are the most popular fastener for most jobs. When the project demands extra strength and a finer appearance, carpenters usually turn to screws or adhesives, or both together. If strength alone is the issue, oversize lag screws or bolts provide the answer.

Fortunately, for special problems there are special fasteners. If toenailing and driving large spikes aren't your favorite pastimes, you may be pleasantly surprised by the many metal framing connectors. And if you're stymied by materials like gypsum wallboard, plaster, masonry, or concrete, read on and you'll find some effective ways to cope.

Nails

Nails are sold in 1, 5, and 50-pound boxes, or loose in bins. As a visit to the nail department of most hardware stores will show, you can choose from scores of types. Here's a guide to selecting the right nail for the job.

The basic nail collection. For most uses, carpenters choose one of these four basic nail types: common, box, finishing, and casing. Common nails come in sizes from 2-penny to 60-penny; box and casing nails from 2-penny to 40-penny; and finishing nails from 2-penny to 20-penny. "Penny" (abbreviated, oddly, as "d") once referred to the cost of 100 hand-forged nails; 16-penny nails, for instance, were 16 cents per hundred. Now the term indicates a nail's length. Here are equivalents in inches for the most common sizes:

4d = 1½''	6d = 2''	8d = 2½''
10d = 3''	16d = 3½''	20d = 4''

- **Common.** Favored for heavy construction, the common nail's extra-thick shank has greater strength than most. A wide, thick head spreads the load and resists pull-through—and makes a good target for a hammer. Common nails more than 6 inches long are called spikes.

- **Box.** Similar in shape and use to the common nail, the box nail has a slimmer shank. These are less likely to split wood, but easy to bend, unless you're experienced with a hammer.

- **Finishing.** When you don't want the head of the nail to show, use a finishing nail. After you drive it nearly flush, sink the slightly rounded head with a nailset (see page 25).

- **Casing.** Similar to the finishing nail, casing nails have a thicker shank and more angular head; use them for heavier work such as adding casings around doors or setting flooring. They can be hard to find.

Special-use nails. Each of the following nail types excels at a particular job. You won't find them all at hardware stores, though; you may have to check building supply shops.

- **Spiral.** A spiraling, grooved, screw-like shank rotates slightly for better holding power as you drive the spiral nail into wood. They won't back out, so they're a good choice for laying and repairing floor boards.

- **Ring-shank (annular-ring).** The closely spaced, grooved shanks provide extra holding power, especially in softer woods. Use them to nail down sheet subflooring or underlayment and to hang gypsum wallboard.

A NAIL FOR ALL REASONS

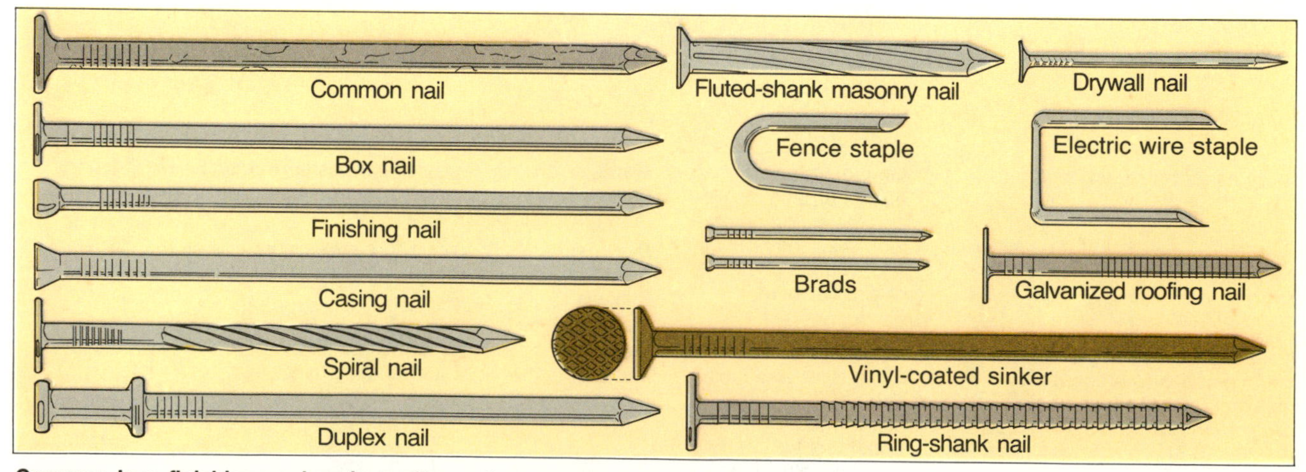

Common, box, finishing, and casing nails are the carpenter's standbys. Other nails come in handy for specific projects.

■ **Cement-coated drywall.** Like ring-shank nails, drywall nails (from 1¼ to 1⅝ inches long) are used to hang gypsum wallboard. The head is slightly beveled on the underside and hollowed on top, making it easier to sink the nails below the wallboard surface.

■ **Galvanized roofing.** The wide heads give roofing nails maximum holding power. There are many variations in size and shape: be sure to buy the nail recommended by the manufacturer of your roofing material.

■ **Fluted-shank masonry.** Useful for securing sole plates for stud walls to concrete, or for fastening furring strips to masonry, these nails are case-hardened for extra strength. Always wear safety goggles when driving them.

■ **Duplex.** This double-headed nail is used for temporary work (like constructing foundation forms or nailing wall bracing). Drive the lower head tight against the surface; pull the nail out by the upper head.

■ **Vinyl-coated sinker.** The coating helps this nail slide easily into older, harder softwoods and hardwoods. (Some carpenters claim that they also come out more easily.)

■ **Cement-coated sinker.** Similar in appearance to the vinyl-coated sinker shown on the facing page, this sinker has a thin coat of cement to provide added holding power.

■ **Brads.** Resembling miniature finishing nails, brads are useful for securing moldings to cabinets and walls or for joining delicate wood edges. They're sized by length and wire gauge: the higher the gauge number, the thinner the brad.

■ **Staples.** U-shaped galvanized staples hold wire fencing to wood posts. Flat-headed electric wire staples are designed to secure nonmetallic cable to wood framing, but they're handy wherever you need a large staple.

Rustproof nails. It's wise to use hot-dipped galvanized, aluminum, or stainless steel nails on the exterior of your house. The best hot-dipped nail will rust in time, particularly at the exposed nail head, where the coating is battered by your hammer. Stainless steel or aluminum nails won't rust, but they cost about three times as much as galvanized nails. They're also hard to find; you'll probably have to special-order them.

Screws

Though more time-consuming to drive than nails, screws create stronger and neater joints—especially when combined with glue.

Screw types. Six kinds of screws commonly used in wood are shown below. The most common screw, the *flathead,* sits flush with the material's surface. The flathead *Phillips screw* is also very popular; a crosslike pattern notched in the Phillips head keeps the screwdriver from slipping.

Roundheads, which sit atop the surface, are used in thin wood or to attach thin material between screw head and surface. The partially recessed *ovalheads* are used for decoration or for attaching exposed hardware. Carpenters and woodworkers also employ *panhead sheet metal screws* for their excellent holding ability.

The sixth type shown, the heavy-duty *lag screw,* is an oversize screw with a square or hexagonal head. Drive lag screws with a wrench or ratchet and socket.

Screw sizes. Lag screws, with shafts from ¼ to ½ inch in diameter, come in lengths from 1 to 12 inches. Woodscrews are sized by length (from ¼ to 4 inches) and, for thickness, by wire gauge number (0 to 20—about 1⁄16 to ⅜ inch). In general, the thicker the wire gauge for a given length of screw, the greater its holding ability.

COMMON SCREWS, COMMON SIZES

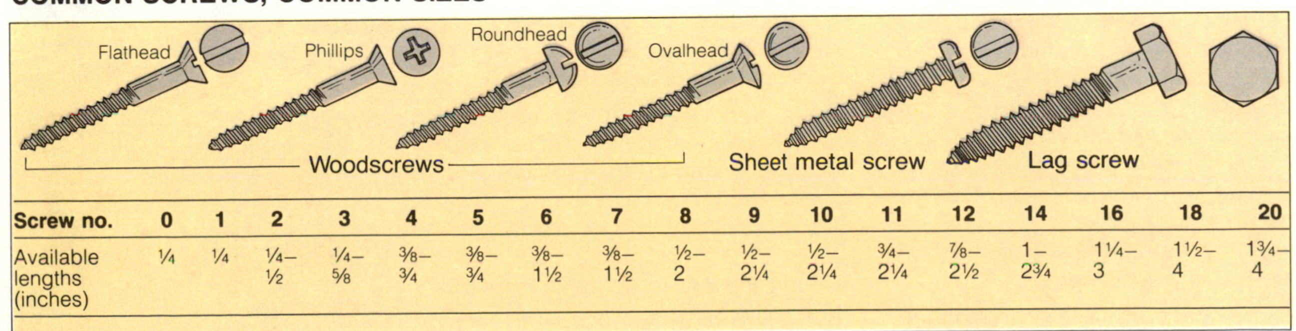

Screw no.	0	1	2	3	4	5	6	7	8	9	10	11	12	14	16	18	20
Available lengths (inches)	¼	¼	¼–½	¼–⅝	⅜–¾	⅜–¾	⅜–1½	⅜–1½	½–2	½–2¼	½–2¼	¾–2¼	⅞–2½	1–2¾	1¼–3	1½–4	1¾–4

This selection of screws, ranging from trim woodscrews to the heavy-duty lag screw, creates strong, neat joints. Standard woodscrew sizes are listed at the bottom.

. . . Fasteners

Bolts

Unlike the screw's tapered point, which digs into wood, a bolt's uniformly threaded shaft passes completely through the materials being joined and is tightened down with a nut. Bolts are stronger than nails or screws because the head and nut grip the material from both sides.

Bolt types. Most bolts are made from zinc-plated steel, but aluminum and brass bolts are also available. The machine bolt's square or hexagonal head is driven with a wrench. Carriage and ribbed bolts have self-anchoring heads that dig into the wood as you tighten the nut. Stove bolts are slotted for screwdrivers.

Bolt sizes. Bolts are classified by diameter (⅛ to 1 inch) and length (⅜ inch and up). To give the nut a firm bite, pick a bolt ½ to 1 inch longer than the combined thicknesses of the pieces to be joined.

If you can't find a bolt long enough for your job, use threaded rod (headless bolt shaft) cut to length, with a nut and washer at each end.

Nuts and washers. Hexagonal and square nuts are the standard, but you'll also see wing nuts, T-nuts, and decorative acorn nuts. T-nuts fit flush against the bottom material, but are weakest.

Most bolts need a flat, round washer at each end. Self-anchoring bolts require only one washer, inside the nut. Lock washers help keep nuts from working loose.

BOLTS, NUTS & WASHERS

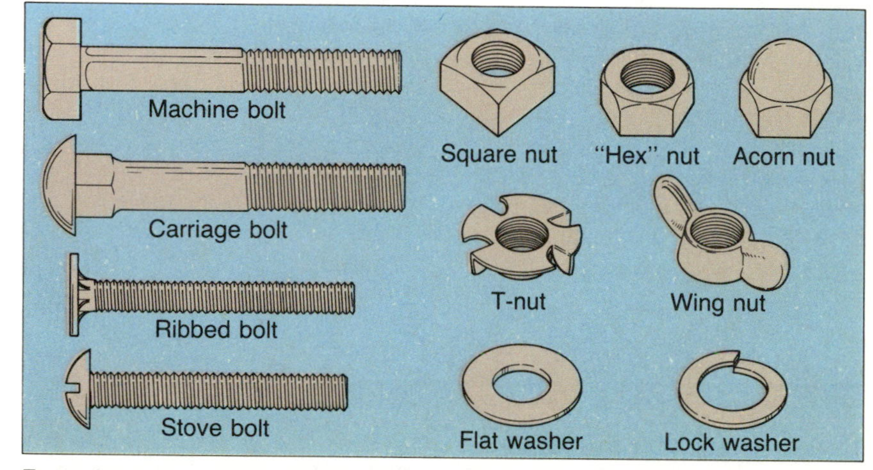

Teamed up with nuts and washers, bolts make the strongest of joints.

Wall and Masonry Fasteners

What do plaster, gypsum wallboard, concrete, and masonry have in common? They lack the resilience necessary to hold common fasteners. So when it's time to hang shelves, picture frames, or mirrors, you'll need special fastening devices.

Wall fasteners. For gypsum or plaster walls, fasteners depend on a spreading frame that distributes weight more widely than a nail or screw. *Spreading anchors,* consisting of a bolt and a metal sleeve, are tapped into a predrilled hole. Tightening the bolt expands the sleeve against the wall's back side. You then back out the bolt, slip it through the object to be attached, and retighten the bolt in the sleeve.

Toggle bolts have spring-loaded, winglike toggles that expand once they're through the wall. Drill a hole large enough for the compressed toggles. Pass the bolt through the fixture to be mounted and attach the toggles; then slide the toggles through the hole—they'll open on the other side and pull up against the back of the wall when you tighten the bolt.

Masonry fasteners. When it's time to secure fixtures or ledger strips to a masonry wall or to anchor a new stud wall to an existing concrete slab, turn to one of these special fasteners.

Lead shields employ hollow-core, threaded sleeves in tandem with woodscrews or stouter lag screws. Drill a hole the diameter of the sleeve and slightly longer, and tap the sleeve in. After slipping the screw through the fixture to be attached, drive it into the sleeve.

More reliable than lead shields, *expanding anchors* feature expanding rings or prongs that grip the surrounding hole firmly when the nut is driven home.

SPECIAL FASTENERS

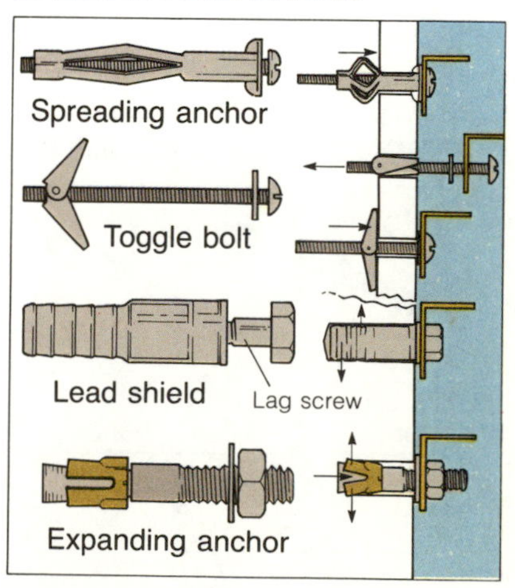

Non-wood surfaces need special fasteners.

Adhesives

Adhesives vary according to strength, water resistance, ability to fill gaps, and setting time. The following list covers those most useful to the carpenter.

White (polyvinyl) glue. The standard household glue, this works well on wood if firmly clamped. Apply it straight from the bottle, and wipe the excess off with a damp rag. White glue isn't waterproof, but it resists grease and solvents. It will soften if used near high heat.

Yellow (aliphatic resin) glue. Often labeled "carpenter's glue," this is a good choice for an all-around adhesive. Though similar to white glue, aliphatic resin has a higher resistance to heat, sets up faster, and is stronger. It can also be applied at temperatures as low as 50° F.

Resorcinol glue. Waterproof resorcinol is used in building boats and outdoor structures, and indoors in wet, humid areas. It leaves a dark stain that may show through paint, so you need to be neat it you're planning a fine finish.

Contact cement. This adhesive bonds immediately on contact and needs no clamps. It's frequently used to attach wood veneers or plastic laminate to wood surfaces. Be careful: since the initial bond is permanent, you must align parts exactly. The older type is highly flammable and noxious; buy the newer, water-base type if you can.

Paneling adhesive. Solvent or water-base paneling adhesive is typically packaged in 11-ounce cartridges and applied with a caulking gun (see page 28). Use it to fasten plywood and hardboard wall paneling, gypsum wallboard, and subflooring materials to framing members.

Metal Framing Connectors

Few things are more frustrating than watching an angled nail split the end of an expensive piece of lumber. And no matter how experienced a carpenter you are, nails alone can't provide the strength that metal connectors can.

Special 1½-inch joist hanger nails are used for nailing connectors to 2-by lumber. Connections to larger pieces (4-by and up) should be made with nails specified by the manufacturer; generally, they require 3½-inch (16d) nails. For outdoor applications, use galvanized or stainless steel nails, if available.

You'll find the following framing connectors—and probably many more—at almost any lumberyard, in sizes to fit either rough or surfaced lumber.

Joist hangers. Probably the most familiar metal connectors, joist hangers make a secure butt joint between a floor or ceiling joist and the load-bearing beam or joist header (see pages 54–61). Some joist hangers have metal prongs that can be hammered into the side of the joist itself. (The connection to a beam must be made with nails.) Joist hanger sizes start with those for 2 by 4s.

Post anchors and caps. Though styles vary, the post anchor is generally used to secure a load-bearing post to a concrete foundation or slab. Attach it to the post with either nails or bolts—you'll see punched holes for them in the sides. In areas where there's much standing water or rain, some carpenters use an elevated post base that allows 1 to 3 inches of clearance above the concrete.

Post caps are used at the top of a post to join the post to a beam and to strengthen a splice connection between two beams.

Framing anchors. Here you'll find a kaleidoscope of connectors, each with its own purpose. Hurricane anchors eliminate toenailing between rafters and top plate; reinforcing angles create solid joints between any two members that cross. Spiked truss ties and top plate ties are simply hammered flat. T-straps and L-straps also make strong connections.

FRAMING CONNECTORS: A SELECTION

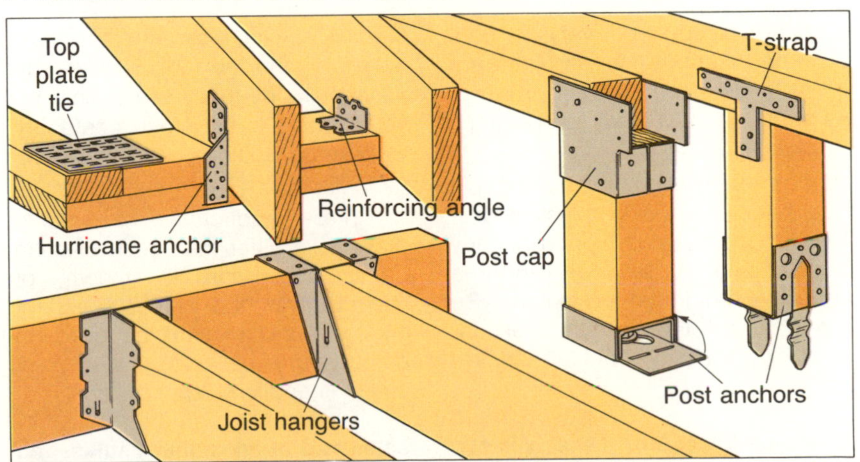

Galvanized steel connectors provide solid joints between framing members like joists, posts, rafters, and beams—and eliminate tricky nailing.

Sealing Products

A house protects its inhabitants from rain, snow, wind, cold, and summer sun. But the structure itself must be able to withstand weather damage over time, and that's where sealing products play a role. Wood preservatives help fight wood decay; they're especially important where wood touches the ground or standing water. Caulking compounds fill exterior cracks or seams that let in damaging moisture, uncomfortable drafts, and shelter-seeking insects. Effective insulation stops warm air from escaping in winter and slows down the accumulation of heat from outside in summer.

These two pages present the materials with which a carpenter ensures a house's ability to keep weather out and comfort in.

Wood Preservatives

Wood preservatives extend the life of lumber by increasing its resistance to the decay caused by fungi, mold, and wood-eating insects.

Pressure treatment, the most effective method of application, forces chemical preservatives deep into wood fibers; lumber from species such as southern pine or Douglas fir can be made as durable as the hardier types—redwood, cedar, or cypress.

Pressure-treated lumber is widely available, though difficult to find in some regions. (Where not available, it can be special-ordered through local building suppliers.) Depending on the amount of chemical injected, the wood may be labeled "22, ground-contact use," or "L.P. 2, above-ground use."

Preservatives can also be applied with a brush or by immersion, but the long-term results are less satisfactory.

The major types of wood preservatives and their most effective uses are outlined below. Because many of these materials are considered toxic, you should read the manufacturer's precautions carefully before applying them. For a discussion of basic safety equipment and procedures, see pages 22–23.

Creosote. The granddaddy of preservatives, creosote is both a water repellent and a preservative. A pressure-treated post or piling may last 40 years or more.

Because creosote has a heavy odor and cannot be covered with paint, its role in residential construction is generally limited to below grade (below ground level).

Pentachlorophenol. This oil-borne preservative is as effective as creosote, except for below-grade use. It's commonly found in commercial water-repellent preservatives. In their clear form, these compounds provide the most effective treatment for woods left in their natural hues. Pentachlorophenol can be applied by brushing, dipping, or soaking. Soaking is the most efficient technique for home application.

Copper naphthenate. Applied by brushing, copper naphthenate is almost as effective as pentachlorophenol. Because it's nontoxic to plants and animals, it's especially useful for treating garden structures. The dark green tinge it leaves on treated wood can be covered with two coats of paint.

Water-borne salt preservatives. Water-borne salts applied to wood are clean and odorless but leave a light yellow, green, or brownish tinge. Highly recommended by experts, their main limitation is that they must be applied by pressure treatment.

Caulking Compounds

Caulking compounds vary in both price and composition, and there's a direct relationship between the two. As a rule, you get what you pay for.

For most caulking jobs, the 11-ounce cartridge and the caulking gun (see page 28) are simplest to use. The surface to which you apply any caulk should be clean, dry, and free from oil and old caulking material. Be sure to check the manufacturer's instructions for any special requirements.

Elastomeric caulks. These synthetic "supercaulks" are the top of the line. They effectively seal almost any type of crack or joint, they adhere to most materials, and they will outlast ordinary caulks by many years. The generic types that fall into this category include polysulfides, polyurethanes, and silicones.

These products do have some drawbacks other than cost. Silicone rubber is awkward to smooth out once applied, and it doesn't accept paint. Polysulfide can't be used on porous surfaces unless a special primer is applied first.

Latex and butyl-rubber caulks. You can also buy all-purpose caulks that offer average performance. These medium-priced products include latex, acrylic latex, and butyl-rubber.

Latex and acrylic latex caulks vary in price and performance. The acrylic latex caulks outperform non-acrylic latexes. Both are easy to apply and clean up with water.

Butyl-rubber caulk, generally more flexible and durable than acrylic latexes, can be used on any type of surface or material. It has a tendency to shrink slightly while curing.

Oil-base caulks. These are the lowest-priced and lowest-performance caulks on the market. Limit their use to stable interior seams and cracks.

Insulation

When winter winds howl and energy bills soar, the well-insulated house pays huge dividends. Here's a guide to the most common homeowner-installed materials and their uses.

Blankets and batts. The popular mineral-wool insulation—fiberglass or rock wool—is most commonly available either in large rolls called blankets or in precut 4 or 8-foot lengths called batts. Both blankets and batts are sized to fit the spacings between wall studs, rafters, or joists.

Loose fill. Loose fill materials are either hand-poured or machine-blown into place. Though these materials are most suitable for unfinished attics, some types are blown into walls that are covered on both sides. Vermiculite and rock wool are the most common hand-poured materials. Cellulose is sometimes hand-poured, but machine installation is much more efficient.

Rigid insulation boards. For exposed-beam ceilings or basement walls, rigid insulation (available in 4 by 8, 4 by 4, or 2 by 8-foot panels) is often the best answer.

How much is enough? Insulation materials are rated by their resistance to heat flow. These ratings are called "R-values." The higher the R-value a material has, the better it will insulate.

RECOMMENDED R-VALUES BY CLIMATE ZONE

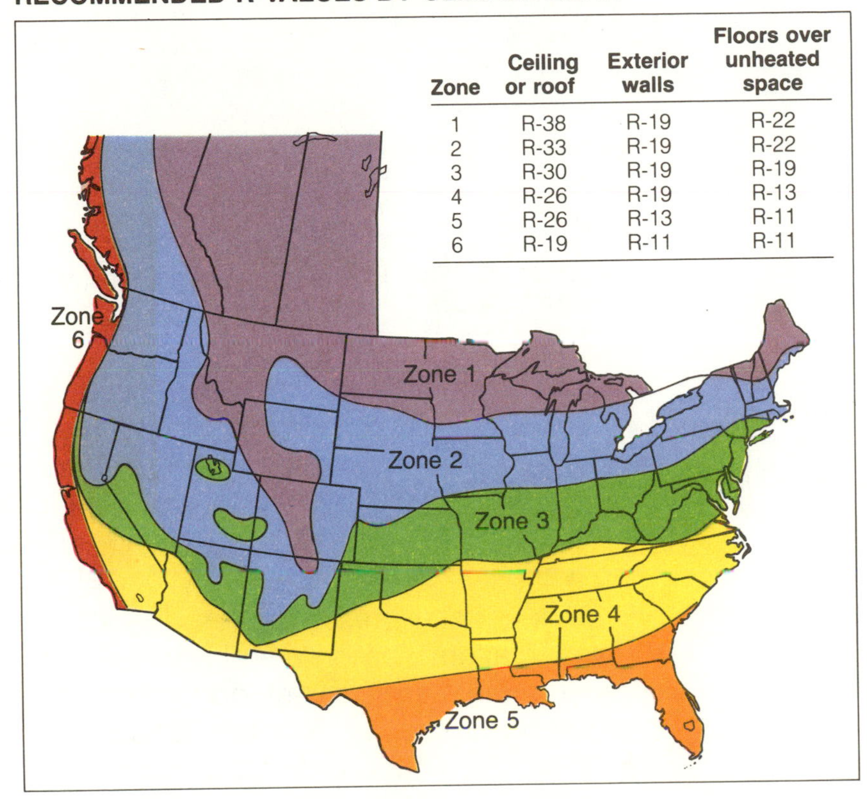

Zone	Ceiling or roof	Exterior walls	Floors over unheated space
1	R-38	R-19	R-22
2	R-33	R-19	R-22
3	R-30	R-19	R-19
4	R-26	R-19	R-13
5	R-26	R-13	R-11
6	R-19	R-11	R-11

Vapor barriers. In all but the driest climates, it's necessary to install a vapor barrier to prevent humid house air from condensing inside insulated walls, floors, and roofing materials. Blankets and batts are commonly sold with a vapor barrier of foil or kraft paper; if your insulation doesn't have this protection, cover it with polyethylene sheeting, foil-backed wallboard, or building paper.

BASIC INSULATION TYPES

Insulation products come in several forms: blankets and batts are most popular; loose fill can be poured where access is limited; some rigid boards double as decorative ceiling panels.

Anatomy of a House

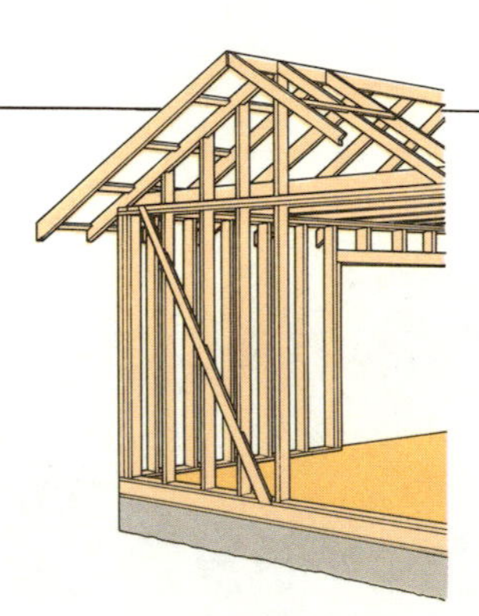

Have you ever secretly wondered, "What's a floor joist?" Or been too embarrassed to ask "What's a wall stud?" If so, you're not alone. But here's your chance to learn about basic building components and see how they fit together. Soon that befuddling snarl of boards by the hundred and nails by the thousand will become a simple, logical system.

This chapter presents three distinct house framing styles: platform, balloon, and post and beam. Platform framing, the standard method for today's house building, is detailed in the last chapter, "Basic Procedures." Here, we'll walk you through an overview of platform framing—from the foundation footings to the trim around the living room window, following the complete sequence you'd take to build it yourself. Next, we'll outline balloon framing, common in older homes, and help you determine whether your house has this type—important information if you're planning to remodel. Finally, we'll discuss the major components and procedures for post and beam framing, a style that's gaining popularity today both for its simplicity and for its architectural flair.

Platform Framing

Platform (sometimes called "western") framing is the simplest, safest house-building method in practice today, and the most common type by far in newer homes. The "platform" consists of the foundation and floor structure; walls are built up from this solid base. If the house is taller than one story, additional layers of floor platforms and walls are stacked atop the first floor walls. Finally, ceiling and roof framing complete the structure.

To explain the basic components and building sequence of a platform-framed house, we're going to "build" one from start to finish. Even if your house building will be confined to daydreaming by the fire, you'll learn how the parts all fit together. And when it's time to take hammer in hand—to hang a picture, fix a squeaky floor, install new wall paneling, or frame a skylight—you'll have a head start.

Start with a Solid Base

A house is only as sturdy as the base on which it rests. That base must support the weight of the entire structure above, distribute the weight evenly to the ground below, and resist lateral movement ("racking").

Foundation. Any foundation begins with wide concrete pads called *footings.* As the name suggests, footings are the platforms on which the legs—*foundation walls* or *posts*—rest.

Though foundation "walls" can be pressure-treated pilings, poles, or surfaced posts, they're normally solid perimeter walls composed of poured concrete or concrete blocks. Depending on the depth of the excavation and foundation walls, these perimeters create a full basement or the shallower "crawlspace."

Atop the foundation walls lie the *mudsills.* The mudsills provide a solid transition from the concrete below to the "2-by" wood framing above and help distribute the house's weight evenly along the foundation.

Most foundation designs include at least one interior wall or a *girder* (supported by posts and piers) that helps bear the load above.

Floor structure. Next come evenly spaced *floor joists,* which rest on opposite mudsills and span the width of the foundation. Where joists meet a girder, they may be continuous, but normally shorter lengths are "overlapped," as shown below. *Joist headers* protect joist ends, help keep the joists upright, and block drafts from the exterior.

If joists span more than 8 feet, it's traditional to insert solid *blocking* or X-patterned wood or metal *bridging* to prevent joists from twisting.

At this point, any necessary below-the-floor work—such as plumbing, heating, wiring, or insulation—is completed.

Finally the *subfloor,* either solid boards or plywood sheets, is laid over the joists.

One variation on the platform design that's popular in frost-free areas is the concrete slab foundation. The continuous slab doubles as both foundation and subfloor, though footings are still required around the perimeters and below interior bearing walls (see page 56). In this type of construction, the walls are attached directly to the slab.

FOUNDATION & SUBFLOOR ELEMENTS

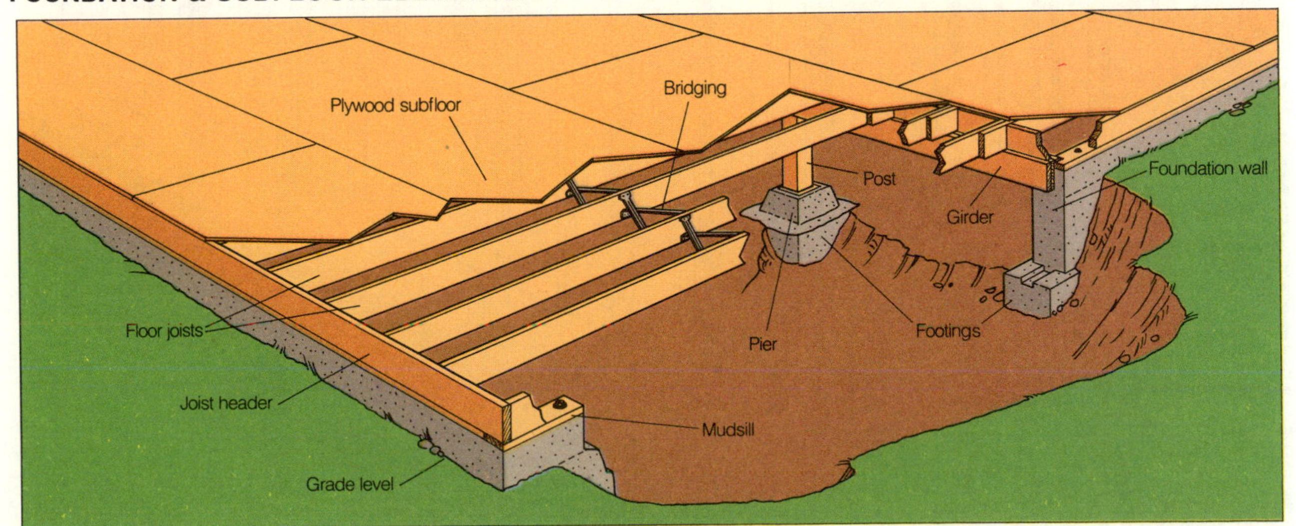

Platform framing begins below grade level with concrete footings and foundation walls. One or more girders lend additional support to the floor joists and subfloor materials above.

...Platform Framing

Frame the Shell

The walls, ceiling, and roof of a platform-framed house are composed almost entirely of "2-by" dimension lumber; nails usually hold together the whole assembly (though increasingly they're supplemented by metal framing connectors).

Walls. The framing job begins with the exterior stud walls, formed by vertical, evenly spaced *wall studs* that run between a horizontal *sole plate* and parallel *top plate.* At window and door openings, *headers* distribute the load normally transferred to the missing studs. *Cripple studs, rough* (window) *sills,* and *trimmer studs* help strengthen openings.

Walls may be either "bearing" or "nonbearing." A bearing wall helps support the weight of the house; a nonbearing wall does not. All exterior walls running perpendicular to ceiling and floor joists, and normally at least one main interior wall positioned over a girder, are bearing.

Where walls intersect, you'll find extra studs or blocking to help tie them together. To prevent racking, exterior corners are frequently braced with 1 by 4 boards or metal straps, though plywood sheathing or siding often makes this unnecessary.

WALL, CEILING & ROOF FRAMING COMPONENTS

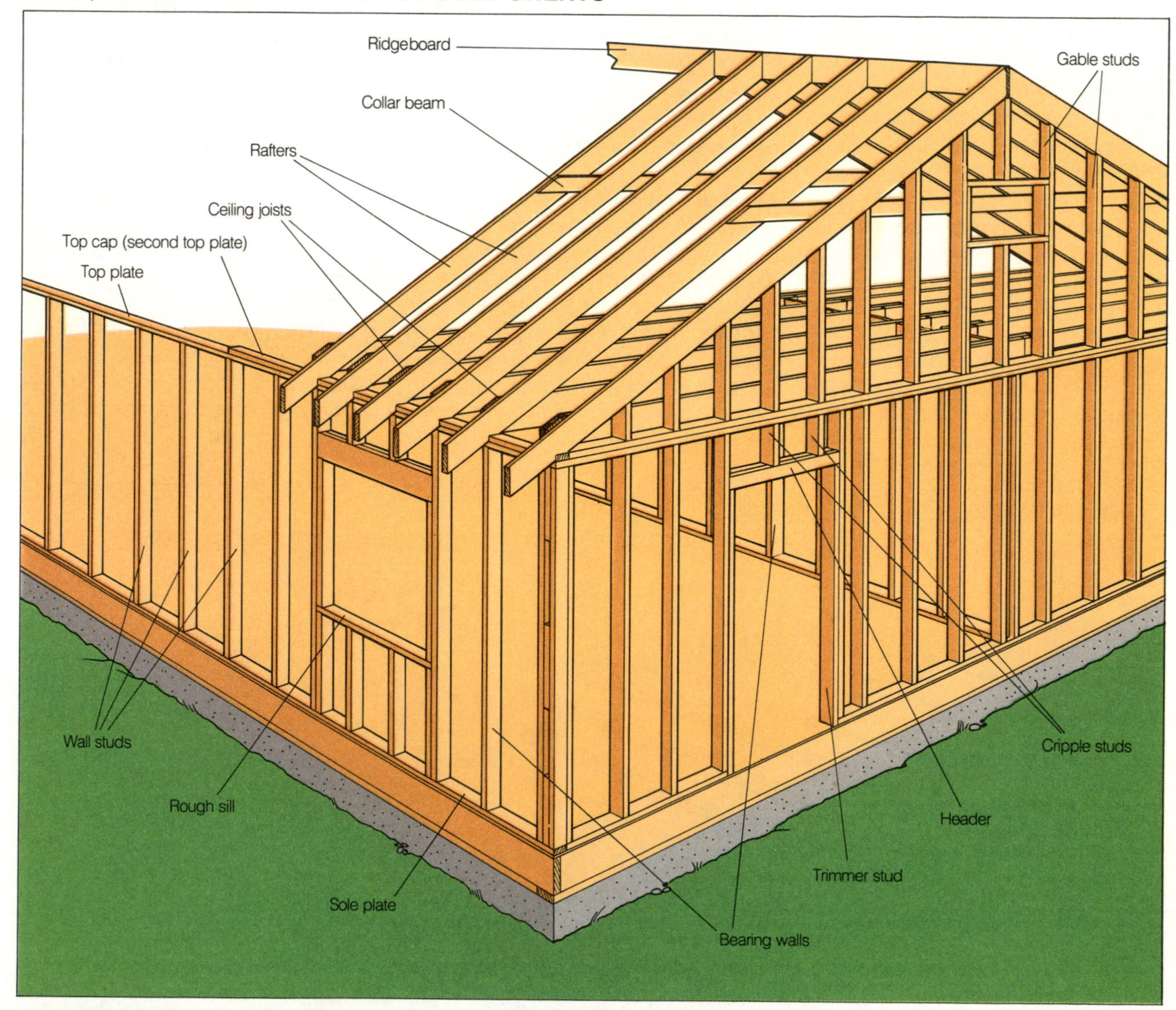

Stud walls, ceiling joists, and rafters complete the house skeleton. The walls are secured to the subfloor and to each other; ceiling joists rest on the top plates, and rafters run from top plates to the ridgeboard at the peak.

The walls are fastened through their sole plates to the subfloor and joists below. A second set of top plates, sometimes called *top caps,* overlaps the first and provides extra rigidity.

Ceiling. Now comes another layer of joists parallel to the floor joists below. If there's no second story, these are lighter *ceiling joists;* otherwise, they'll be another set of heavy-duty floor joists. In the drawing on page 56, note how the interior bearing wall supports the joists above as the girder did below.

Roof. The framing for a straightforward gable roof consists of matching pairs of evenly spaced *rafters* that meet at a central *ridgeboard.* The rafters are notched to fit over the top plates; the distance the rafters extend beyond the plates determines the roof overhang or eaves. The ends of the ceiling joists are trimmed to match the rafter slope.

Rafters that span a long distance are often bridged with horizontal *collar beams* or *ties,* or supported by horizontal *purlins* and braces (see page 79). At the open ends, vertical *gable studs* are angled to meet the rafters above the top plate.

An increasingly popular alternative to standard roof framing involves engineered *trusses,* which combine ceiling joist, rafters, and diagonal bracing into single, triangular units. Though roofs go up fast this way, you'll need heavy machinery to raise trusses atop the walls.

Apply the Exterior Skin

Once the house is framed, the immediate objective is to protect it from the elements by enclosing the framing.

Sheathing. Walls may require a layer of *sheathing* to help strengthen the structure and to provide a nailing base for siding materials. (Plywood sheet siding may not require sheathing.) A layer of *building paper* covers the sheathing or wall studs.

Roofs also need a deck of sheathing to form a base for the finish roofing materials. Depending on your roofing choice, sheathing can be solid boards or plywood or open, spaced boards. A layer of *roofing felt* often goes atop the deck.

Finishing touches. Next, it's time to secure windows and exterior doors to their framed openings, adding door sills and thresholds as necessary. With this task complete, the finish siding and roofing materials go on.

Exterior finish work may include trim at house corners, *fascia* boards attached to rafter ends at the eaves and along gables, or enclosed *soffits* (see page 85) below the eaves. Adding exterior window and door *casings* as needed completes the exterior.

ROOFING, SIDING & TRIM

Seal the house framing from the elements with wall and roof sheathing, windows and doors, and finish siding and roofing materials. With the exterior trim complete, it's time to move indoors.

...Platform Framing

Close Up the Interior

Now your house is "built" and the rains may rage. But after the well-earned celebration, a whole new phase of carpentry begins.

Roughing-in. Once the work moves indoors, it's time for the plumber, electrician, and heating specialist—or the energetic do-it-yourselfer—to rough-in pipes, electrical cable and boxes, and the heating system. It's also time to frame a set of basement stairs, perhaps, or a walk-in closet.

Insulation comes next: our model shows mineral-fiber blankets installed between exterior wall studs. If the attic area will remain unfinished, also place insulation between ceiling joists; insulate finished attics between rafters and gable studs. A vapor barrier, often included with insulation materials, is positioned between the main living areas and the insulation.

Finish work. First, ceiling materials are fastened directly to the ceiling joists or to a metal or wooden grid suspended from the joists. Gypsum wallboard or plywood sheet paneling is applied directly to wall studs; solid board paneling may first require a wallboard backing, or furring strips as a nailing base. Plaster walls need a gypsum, fiberboard, or metal lath backing. If you choose wallboard as a final covering for either ceilings or walls, you'll need to finish, then sand, the joints between panels.

If the finish floor covering requires a separate underlayment—plywood, particleboard, or hardboard—over the subfloor, that's the next step. After the finish floor is down, install interior doors in their openings. Interior trim—baseboards, moldings, and window or door casings as needed—completes the basic job.

Of course, there's still work to be done, depending on the complexity of the job. Paint and wallpaper head the list; cabinets and countertops are next in line. Plumbing fixtures must be hooked up to rough supply, waste, and vent pipes; appliances require gas, water, and electrical connections. Finally, it's time to install light fixtures and electrical cover plates.

There, at last, you have it: the modern platform-framed house.

INTERIOR FINISH WORK

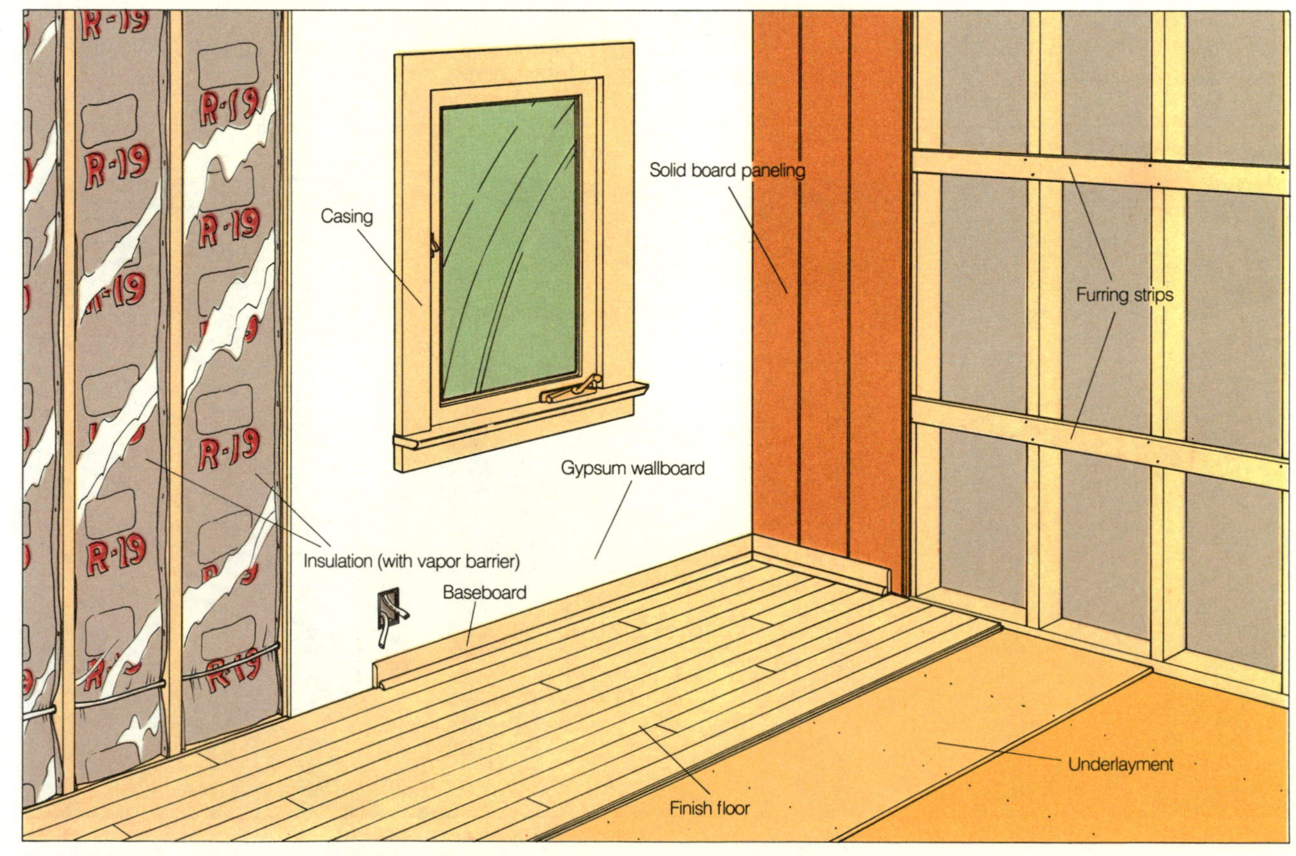

To complete the interior, first rough-in utilities and install insulation; then add ceiling and wall coverings, interior doors, finish floors, and trim.

Balloon Framing

Standard practice until about 1930, balloon framing is still used in some two-story houses, especially those with stucco, brick, or other masonry exteriors. If you're planning a remodeling project that involves cutting into wall framing, you'll need to know if that wall has a balloon frame.

Major components. Like the platform frame we've discussed so far, the backbone of a balloon-framed house includes footings and foundation walls, floor joists, subfloor, wall studs, ceiling joists, and rafters. Wall sheathing and roof decking complete the structural skin.

The chief difference between the platform frame and the balloon frame shown at right is that, in balloon framing, the bearing wall studs extend continuously from the mudsill on top of the foundation wall to the doubled top plate two stories above. Both studs and floor joists are nailed to the mudsill and then nailed to each other. At the second-story level, joists rest on a *ribbon* or ledger strip notched into the studs, and are fastened to the studs as well.

Because the continuous stud spaces form natural chimneys that create a fire hazard, *fireblocks* are required between studs at each level. If positioned carefully, the blocks form a firm nailing base for subflooring or wall paneling.

Interior bearing wall studs normally run one story only—similar to platform framing but with a single top plate. Second-story joists and studs rest directly on the plate. If plaster and lath will cover the wall, though, it may be built with continuous studs to reduce shrinkage.

Diagnosing your present house. How can you tell if an existing structure has balloon framing? Go down into the basement or crawlspace and look for the mudsill atop the foundation wall. If you see paired joists and studs resting on the sill, your house was built with balloon framing. If joists alone are visible below the subfloor, you have platform framing.

Advantages and disadvantages. When constructing a two-story house, balloon framing has two possible advantages over platform framing. As already mentioned, the continuous studs create a more stable surface for masonry sidings and wall coverings. Wall shrinkage and settling are more uniform because only half the number of studs are used. In addition, skilled carpenters can erect the walls in less time than is required for platform framing.

One real drawback, though, is the hazard of working at heights without the solid subfloor of platform framing. Another is the difficulty of finding quality 18 to 20-foot studs.

BALLOON FRAMING COMPONENTS

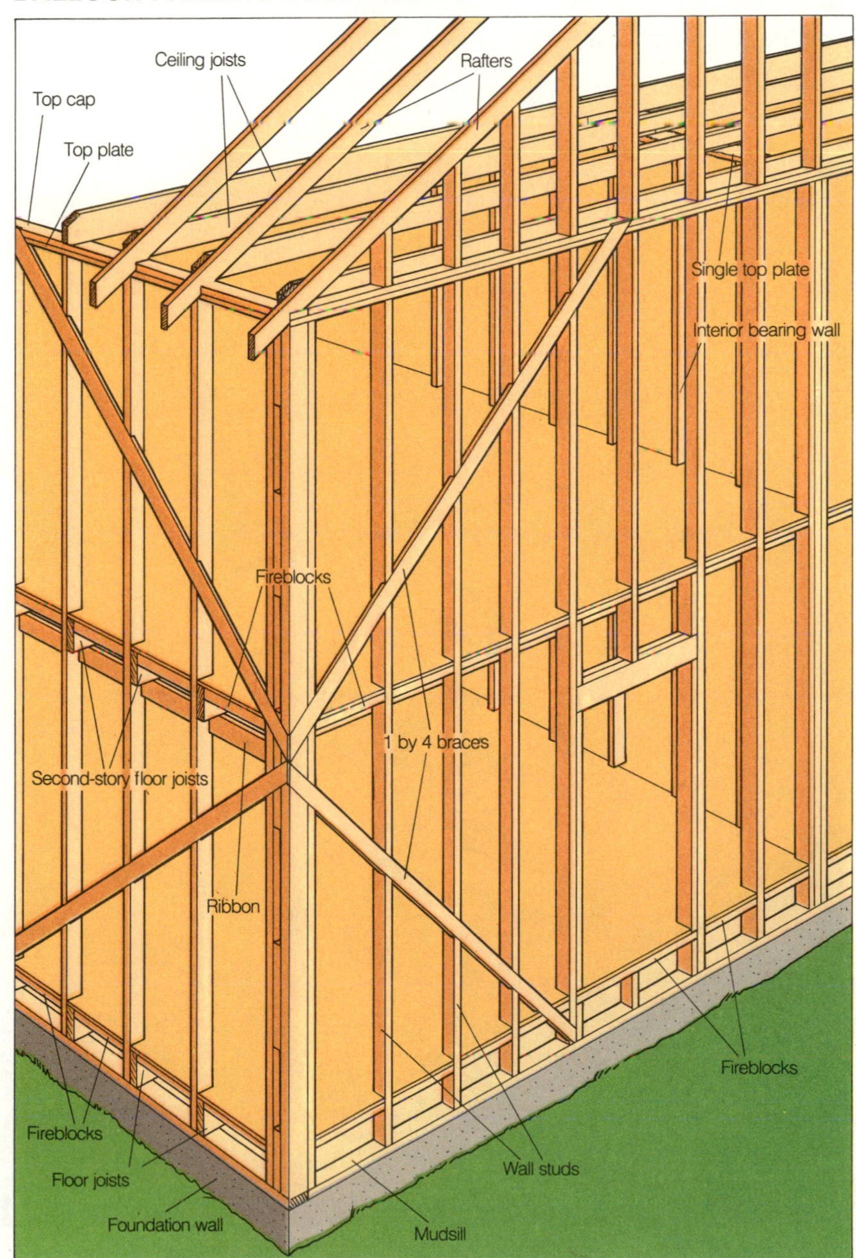

Common in older, two-story houses, balloon framing features exterior studs that extend in continuous lengths from the mudsills to the top plate two stories above.

Post & Beam Framing

The major complaint that designers and builders voice against platform framing is its tendency toward monotony. Post and beam framing (more accurately called "post, beam, and plank" framing) not only lends variety to architectural styles but can help keep both material and labor costs lower than standard construction.

The post and beam design is an updated version of traditional timber and pole framing styles, which utilized heavy structural members at long intervals instead of light 2-by dimension lumber closely spaced.

The basic frame. Post and beam houses consist of sturdy *posts* (up to 8 feet apart) that hold up the *beams*, which in turn support the *planks* forming the floor, ceiling, and roof deck. One trademark of the system, the open-beam ceiling, shows off the clean, bold lines of the planks and beams. The design also lends itself to large expanses of glass, since the wide spaces between posts don't bear any weight.

The heavy-duty posts—a minimum of 4 by 4 surfaced lumber—either sit atop a standard foundation wall and footings or extend down to individual piers and footings.

Beams may be either transverse (much like heavy-duty rafters) or longitudinal (parallel to the length of the structure), allowing for great flexibility in roof profiles.

POST & BEAM FRAMING

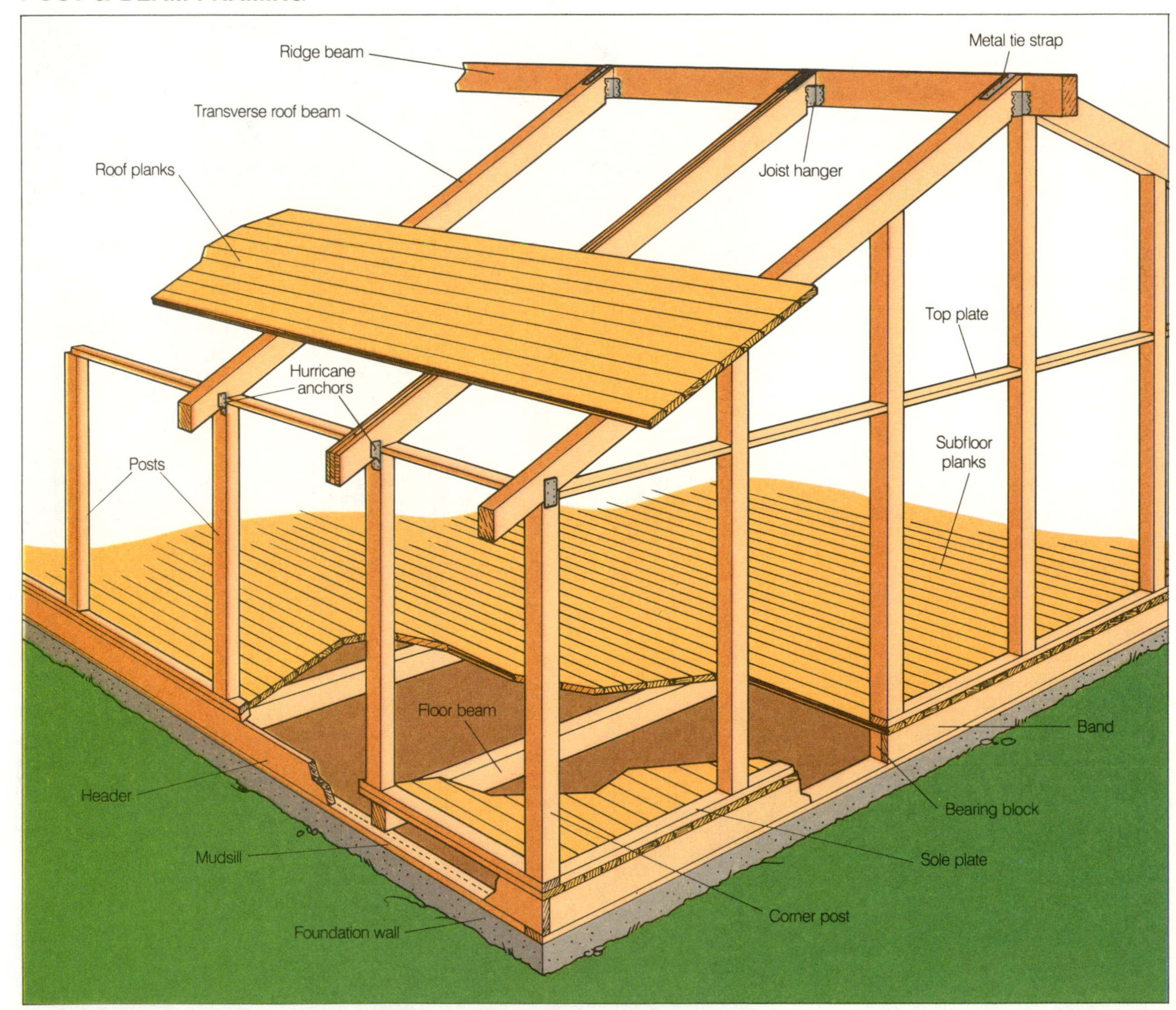

Posts, beams, and planks compose the backbone of a "post and beam" house. The sturdy, vertical posts and horizontal beams allow greater spacing between framing members; planks tie the structure together and serve as subfloor, ceiling, and roof deck.

Both posts and beams can be solid lumber, manufactured glue-laminated members, or "built-up" 2-by sections nailed together carefully.

The 2-inch-thick planks, commonly locked together with tongue-and-groove edges, tie the structure together. Planks should span three or more beams for solid support.

Are there disadvantages? The sheer weight of heavy posts and beams makes them difficult—and potentially dangerous—for do-it-yourselfers to manipulate. Also, because each post and beam carries a large load, the house design must be precisely engineered and framing connections carefully made. Ordinary nails won't do the trick; metal framing connectors or tie straps, secured with lag screws or bolts, are needed. While their appearance can be objectionable, they can normally be concealed from sight.

It's more difficult to conceal overhead electrical and plumbing lines. By installing "spacer blocks" between doubled beams and covering the gap at the bottom with trim, you can form inner passageways where the lines can be hidden from view.

Filling in the walls. In one-story structures, it's common to run a continuous sole plate and top plate from post to post, as shown on the facing page. The spaces between posts may then be filled in with light dimension lumber to provide a framework for exterior and interior paneling, insulation, and electrical or plumbing lines.

If you opt for large amounts of glass or many doors, note that headers are not required, since these wall sections bear no weight. Diagonal bracing or a sufficient number of solid panels must be used, though, to ensure that the house can adequately resist racking.

Interior walls are normally nonbearing. If these partitions run parallel to the floor beams, no support is necessary below: the wall's weight will be spread across the planks above. If the wall is perpendicular to the beams, an extra-heavy sole plate or a beam directly below the wall distributes the load to the main beams. A bearing wall must be placed over a beam.

Laying the roof. When roof planks serve directly as the ceiling below, it's difficult to insulate. One solution: Lay rigid board insulation over the deck and then install plywood sheathing atop the insulation as a nailing base for roofing materials.

FILLING IN THE WALLS

To complete a post and beam structure, roofing materials are laid atop the planks; the spaces between posts are filled with light stud framing and siding, glass, or doors. If an interior wall runs parallel to planks, a heavy sole plate or support beam below is required.

Basic Procedures

Whether you're replacing crumbling wall paneling or raising a new structure from ground to roof peak, this chapter is designed to help prepare you for the task. You can learn about the entire building process—beginning with floor framing and ending with the trim around a new window—and apply those basic procedures to your own plans. If you're adding only a few improvements, turn directly to the appropriate sections.

You'll often hear carpenters talk of "rough" and "finish" work. Rough carpentry means framing the floors, walls, and roof, and adding siding and roofing materials. The more refined finish work focuses on doors, windows, ceilings, stairways, wall coverings, and trim. Just as you would in an actual project, we start here with rough carpentry and then move into finish work. Special features along the way introduce you to the art of reading plans, installing energy-efficient insulation, and adding a new skylight.

Are you the plumber, electrician, or mason, too? The *Sunset* books *Basic Plumbing Illustrated, Basic Home Wiring Illustrated,* and *Basic Masonry Illustrated* can provide help.

Before You Begin

Before plunging headfirst into any carpentry project, carefully consider what's involved. Your checklist should include investigating local restrictions and building codes; deciding whether or not you need help; selecting and purchasing materials; and thinking through the exact sequence of building steps.

Deed and zoning restrictions. Read your property deed before embarking on any building or remodeling project. It may restrict the style of architecture, materials used, or even colors of exterior paint. The deed may also specify where an addition can be placed, and it may restrict the construction of garages or second stories.

Zoning ordinances define the type of occupancy, sewage regulations, and the use of municipal water. These ordinances also regulate the maximum percentage of land coverage; front, side, and back property setbacks; and the heights of fences and utility buildings.

Building codes. Most government building departments have adopted one of the four national model codes. Throughout the western states, this means the *Uniform Building Code*. The *Standard Building Code* is the rule in southern states; the *Basic Building Code* is found in the East and Midwest. The *National Building Code* also affects some areas.

State, county, or local building departments may only adopt part of one of these codes, though; rewriting or adding specifications to meet their own needs. Your project's design and materials must measure up to your particular local code.

Building codes can be exasperating in their breadth and meticulousness. Two condensed versions by the designers of model codes are available in booklet form: "Dwelling Construction under the Uniform Building Code," and the Basic Building Code's "One and Two-Family Dwelling Code," which also lists relevant plumbing, mechanical, and electrical codes.

Building permits. When do you need a permit? According to most codes, a permit is required for almost every structural job you can think of. In actual practice, there's a little more leeway. Though codes vary, they're seldom concerned with cabinetry, paneling, painting, and other projects that do not "alter the use" of the basic structure.

If you are doing all the work yourself, discuss your ideas with the building inspector and ask whether you'll need an application for a building permit. If you do, you'll have to describe the planned work and give the lot, block, tract, street address, or assessor's parcel number of the location; indicate use or occupancy; state the valuation; and submit two to four sets of building plans. In earthquake and slide areas you may also be required to file a geologist's report. Depending on the complexity of your project, you may have to take out an electrical and plumbing permit in addition to the building permit.

Hiring help. For a small project, you can solicit the help of family or friends when you need an extra set of hands. But when building a sizable structure or beginning a major remodel, you may need to hire some help.

Workers earning more than a certain minimum wage must be covered by worker's compensation insurance to cover possible job-related injuries. If they are hired by a licensed contractor, they're covered by the contractor's policies. But for those whom you pay directly, you must carry the insurance.

For an extensive project that will involve many hours of labor, you may also need to register with state and federal governments as an employer, withhold and remit income taxes and disability insurance, and pay social security and unemployment costs.

Subcontractors can prove invaluable when there are specialized tasks to be done. They can offer you expert and efficient work in trenching, foundations, plumbing and heating, electrical work, and so forth.

If you decide to work with subcontractors, you're the boss. On a large job, this can become a seemingly endless task—coordinating all the work, arranging for permits and inspections, scheduling deliveries, and paying the subcontractors. If you can't spare the time and energy, consider hiring a general contractor to oversee part—or all—of the project. Work out an agreement that allows you to perform the jobs you wish to do yourself; itemize specific points in a contract that you both sign.

Shopping for materials. First estimate your materials (for basic guidelines, see page 39). Then make a down-to-the-last-nail shopping list.

Remember the three basic rules of cutting materials costs: 1) order as many materials as possible at a single time from a single supplier; 2) choose your supplier on the basis of competitive bids from several retailers; 3) order materials in regularly available, standard dimensions and in quantities 5 to 10 percent greater than your estimated needs.

If part of the construction is being done by a licensed contractor, he or she may arrange to purchase materials for you at a professional discount. Be on the lookout for usable materials salvaged from dismantled structures. In some areas, wrecking companies specialize in the sale of such salvage. Contractors with large projects nearing completion are also good sources. Be sure to check local codes before using salvaged materials.

Having trouble with a materials list? Take your plans with you to the lumberyard; often the people there can help you prepare a detailed list.

Make a schedule. To avoid perplexing problems later, always think through in advance the exact sequence of building procedures you plan to follow. Ask yourself how much time the project will take. *Always* allow extra time. Plan work days carefully so you're not caught by nightfall or threatening weather when you've just opened up your home's exterior for a new window or skylight!

Floor Framing

As soon as your new structure's foundation has sufficiently cured, you can begin the four-phase process of framing the floor: 1) installing the mudsills along the tops of the foundation walls; 2) assembling any girders and supporting posts that span the interior of the crawlspace or basement; 3) installing floor joists, joist headers, and blocking or bridging between joists—all of which make up a frame for the subfloor; and 4) laying a plywood or solid board subfloor atop the joists.

Floor framing, though structurally simple, must be done with great care—if the finished platform is not level and square, these imperfections will plague you throughout the entire building sequence.

Of course, if your foundation is a concrete slab, you won't need this framing; you can turn directly to page 69 and begin assembling the exterior walls.

Beginnings: The Mudsills

The mudsills—"sills" for short—lie face up on the foundation walls, flush with the outside edge of each wall. Sills are typically made from 2 by 6 lumber; building codes often specify commercially pressure-treated or naturally decay-resistant woods such as all-heart redwood.

Check foundation for square. The best of foundation walls may be slightly askew in the corners or a trifle long on one side. Rather than follow these flaws, it's important to lay the sills square. To check square, measure the diagonals between opposite corners, as shown below; the lengths of AD and BC must match exactly. Also check the lengths of the foundation walls against your planned dimensions.

If your measurements don't agree, you'll need to mark the offending corners—such as "long ½ inch" or "short ¼ inch"—to establish the correct corners. Now measure in from each corner—adding or subtracting for any discrepancies—the width of the sill material (approximately 5½ inches for a 2 by 6 sill). Snap a chalkline between each pair of marks to indicate the inside edge of each sill.

Drill holes for anchor bolts. Choose the longest, straightest sill material you can find. Hold the first piece against the anchor bolts embedded in the foundation wall and transfer the location of each bolt to the sill (see drawing below). Also measure the distance from the chalkline to each bolt's center. Remove the sill and drill slightly oversize holes (about ¾ inch for a standard ½-inch-diameter anchor bolt). Repeat this procedure for each length of sill material you'll use.

Termite barrier, sill sealer. If you're in termite territory, you may need to install a termite barrier between the foundation wall and the sill. Termite barriers must be a minimum of 26-gauge galvanized iron or another acceptable metal. Termite barrier or no, protect the gap between foundation wall and sill with fiberglass sill sealer, available in 50-foot rolls.

Installing the sills. Position the sills on the anchor bolts, install washers, and screw the nuts down fingertight. Now, using a water level, check the sills for level all around the perimeter.

Sills level? If so, simply recheck the diagonals, make final adjustments, and tighten the nuts. If not, you'll need to level them with *shims*—tapering lengths of cedar shingles—driven in from both sides every 4 feet or so along a problem wall.

THREE STEPS TO A STRAIGHT SILL

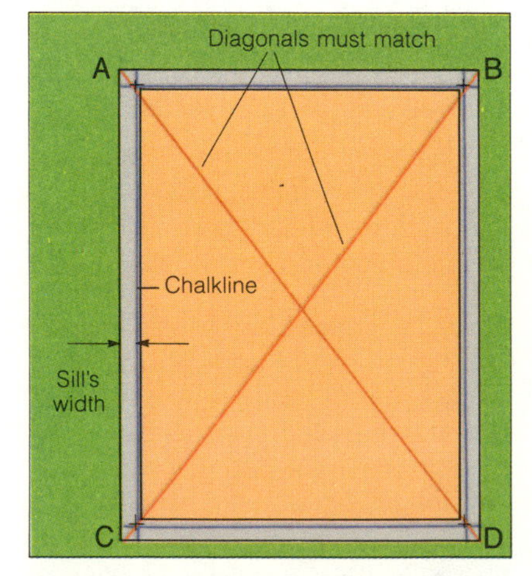

Check the foundation walls for square; diagonals AD and BC must match. Then mark the sill's inside edge at each corner and connect the points with a chalkline.

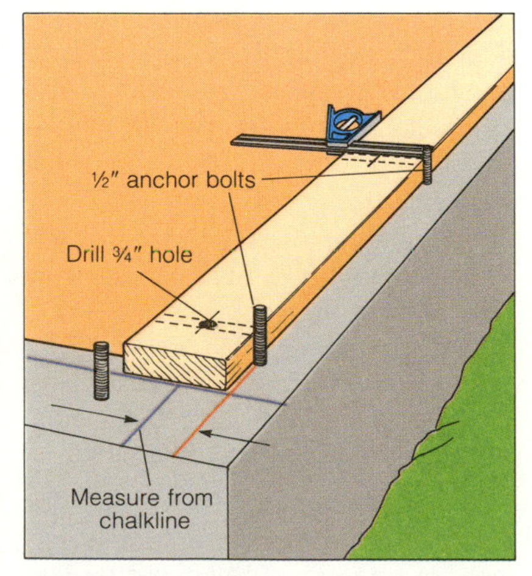

Transfer the outline of each anchor bolt to the sill material; also measure the distance from the chalkline to each bolt center. Drill oversize holes in the sill.

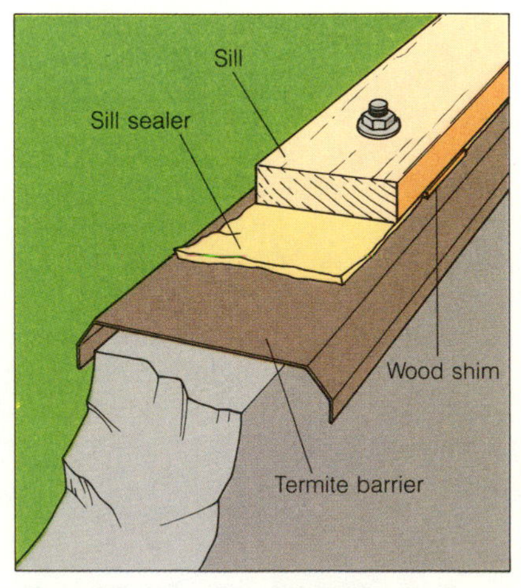

In termite country, a termite barrier is essential. Next come a fiberglass sill sealer and the sill. To correct a sloping sill, use wood shims.

Girders and Posts

A structure wider than about 16 feet needs additional support both for the floor joists at midspan and for any interior bearing walls (see page 69) above. Traditionally, that's a job for a girder. The girder runs the length of the building, parallel to the exterior bearing walls. It's supported in turn by the foundation walls and by wooden posts or steel columns resting on interior concrete footings.

Design notes. Girders may be steel beams, solid lumber, or "built-up" lengths of 2-by dimension lumber nailed together. A built-up girder is the simplest to handle, since it's assembled while it rests on the foundation walls near its final destination.

Posts are either solid lumber, built-up lumber, or concrete-filled steel columns (known as lally tubes). The standard solid lumber post sits atop a concrete pier or metal post anchor embedded in the footing.

The design of girders and posts is interrelated: the depth and thickness of the girder is determined by its span, the size and placement of posts below, and the load of the structure above—all details specified by your building department.

In many areas, wood posts—and girders within 12 inches of the ground—must be pressure-treated or naturally decay-resistant lumber.

Installing the system. The simplest way to assemble a post and girder structure is to position the posts first and then lay the girder right on top.

Using mason's twine, string a line even with the top of the mudsills, designating the center of the girder (see drawing above right). Check level with a water or line level.

Now measure the distance between each pier or post anchor and the string; subtract the depth of the girder and cut the posts to length. Brace posts higher than 12 inches by driving 2 by 4 stakes into the ground, then tacking a 1 by 3 brace to each stake with one duplex nail. Center each post below the string and check it for plumb, using a carpenter's level on adjacent sides. Finish nailing the braces; then toenail the post to the pier's bearing pad or screw it to the post anchor. If post caps are used, they must be set now before the girder is positioned.

The typical built-up girder consists of three thicknesses of 2-by lumber. Nail the pieces together with 20-penny nails spaced 32 inches apart along the top and bottom edges. If you must use shorter lengths to assemble a long girder, stagger the joints between successive layers; when the girder is installed, these joints must be located over a post. Be sure the crowns on the pieces are aligned on the same side. (The crown is the "high" side of an edgeline warp or crook—see page 38.)

With as many helpers as you can recruit, place the girder (crown side up) in its foundation wall pockets (see bottom drawing below). To raise a girder level with the sill, use metal shims—they won't compress or rot. Toenail the girder to each post with 10-penny nails; or, if you're using post caps, nail or screw them to the girder.

ASSEMBLING POSTS & GIRDERS

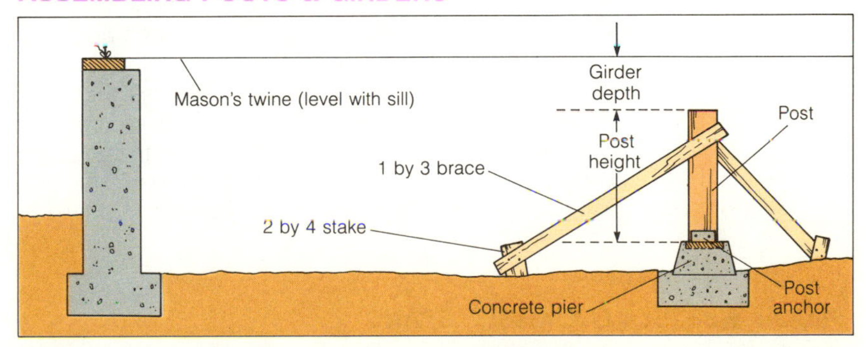

To figure post heights, string mason's twine level with the sill tops. Measure from the pier or post anchor to the twine, then subtract the girder's depth. Fasten each post to its pier or anchor, check plumb, and brace.

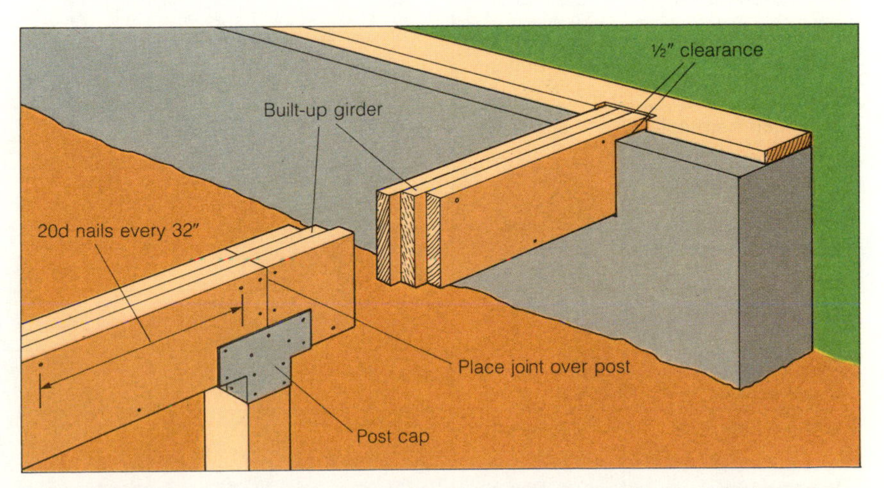

Build a girder from 2-by lumber and position it atop the posts with its ends resting in foundation wall pockets. Toenail the girder to the posts or use metal post caps.

. . . Floor Framing

Floor Joists

Floor joists, evenly spaced lengths of 2-by lumber set on edge, run the width—or short dimension—of the building, spanning opposite sills or resting at one end on a center girder. At the sill, joists butt against a continuous joist header that sits flush with the sill's outside edge. Blocking or bridging is fitted between joists at regular intervals to keep twisting at a minimum and to keep joists in line.

Design notes. Common joist sizes are 2 by 8, 2 by 10, and 2 by 12. The size you choose depends on the distance spanned, the load to be carried, the lumber species and grade you're planning to use, and the spacing between joists. (Framing members are evenly spaced, and the distance between them is usually specified in inches "on center" (*O.C.*)—the measurement from the center of one member to the center of the next.) The chart below lists maximum spans for several common species and grades. Be sure to check the requirements in your area.

Where joists cross a girder, they may continue on, butt together, or overlap. Overlapping allows you to buy and use shorter joists.

Joists are doubled below a concentrated load (such as a cast iron bathtub), around floor openings, and below interior partition walls. Where plumbing lines run up through a partition, doubled joists are often spaced apart with blocks.

Marking the joist layout. First, lay out the joist spacing on one sill, as shown on page 67. Starting at an outside corner, mark the 1½-inch thickness of the first joist, or *stringer*, with a line and an X. For the standard 16-inch joist centers, hook your tape measure over the sill's end and measure off 15¼ and 16¾ inches; then mark another set of lines and an X between. From these lines, continue in even 16-inch intervals to the far end of the sill. When you're doubling joists, mark another line to one side, but don't disturb the basic 16-inch spacing. Don't worry if the last interval is smaller than 16 inches.

Once the layout is complete, transfer the same spacing to the opposite mudsill or the girder. (One exception: If you're overlapping joists, the layout on the opposite sill must be offset 1½ inches to allow for the overlap. Mark the outline of the first joist at 13¾ and 15¼ inches from the end; then mark every 16 inches as before.)

Installing the joists. Choose the straightest, driest lumber you can find for floor joists and joist headers. Sighting along each joist, find the crown and mark that side; install the joists crown side up.

To install the joists, begin by toenailing the joist header flush with the sill, as shown on the facing page, using 8-penny common or vinyl-coated sinkers every 16 inches. Cut the floor joists to length (1½ inches short of the far end), butting them against the header; attach them with three 16-

FLOOR JOISTS: ALLOWABLE SPANS

Species or species group	Grade	Span (feet and inches)					
		2 by 8		2 by 10		2 by 12	
		16" O.C.	24" O.C.	16" O.C.	24" O.C.	16" O.C.	24" O.C.
Douglas fir/larch	2 & Better	13'1"	11'3"	16'9"	14'5"	20'4"	17'6"
	3	10'7"	8'8"	13'6"	11'0"	16'5"	13'5"
Hem/fir	2 & Better	12'3"	10'0"	15'8"	12'10"	19'1"	15'7"
	3	9'5"	7'8"	12'0"	9'10"	14'7"	11'11"
Mountain hemlock	2 & Better	11'4"	9'11"	14'6"	12'8"	17'7"	15'4"
	3	9'7"	7'10"	12'3"	10'0"	14'11"	12'2"
Western hemlock	2 & Better	12'3"	10'6"	15'8"	13'4"	19'1"	16'3"
	3	9'11"	8'1"	12'8"	10'4"	15'5"	12'7"
Engelmann spruce	2 & Better	11'2"	9'1"	14'3"	11'7"	17'3"	14'2"
	3	8'6"	6'11"	10'10"	8'10"	13'2"	10'9"
Lodgepole pine	2 & Better	11'8"	9'7"	14'11"	12'3"	18'1"	14'11"
	3	9'1"	7'5"	11'7"	9'5"	14'1"	11'6"
Ponderosa pine & Sugar pine	2 & Better	11'4"	9'3"	14'5"	11'9"	17'7"	14'4"
	3	8'8"	7'1"	11'1"	9'1"	13'6"	11'0"
Idaho white pine	2 & Better	11'0"	9'0"	14'0"	11'6"	17'1"	14'0"
	3	8'6"	6'11"	10'10"	8'10"	13'2"	10'9"
Western cedars	2 & Better	11'0"	9'7"	14'0"	12'3"	17'0"	14'11"
	3	9'1"	7'5"	11'6"	9'5"	14'0"	11'6"

Design Criteria: Strength—10 pounds per square foot "dead" load plus 40 pounds per square foot "live" load.
Chart courtesy Western Wood Products Association.

penny nails. If your toenailing technique is a bit shaky, attach several joists to the header first; then pull the header flush with the sill and toenail—the weight of the joists will help keep the header from moving. Once the joists are in place, also toenail each to the sill with three 8-penny nails.

At the far end, nail the header to the joist ends and toenail the header to the sill. Where joist ends rest on a girder, joists should overlap a minimum of 4 inches. Tie the joists together with 10-penny nails and toenail them to the girder.

Framing an opening. You'll need to add trimmers, headers, and tail joists around any floor opening you're planning. If the opening is more than 4 feet wide, use double headers. Joist hangers should be used to attach headers longer than 6 feet to the trimmers, and tail joists longer than 12 feet to the headers.

When assembling the framing, temporarily omit the full-length joist on each side to allow yourself nailing room. First, install the initial trimmer joist on each side of the opening. Next, attach the first headers with three 16-penny nails on each end. Cut the tail joists to length and nail them with 16-penny nails, too. Now add second headers inside the first, nailing them through the trimmers as before, and face-nail them to the first headers. Finally, double the trimmers, nailing them together along top and bottom edges with 16-penny nails.

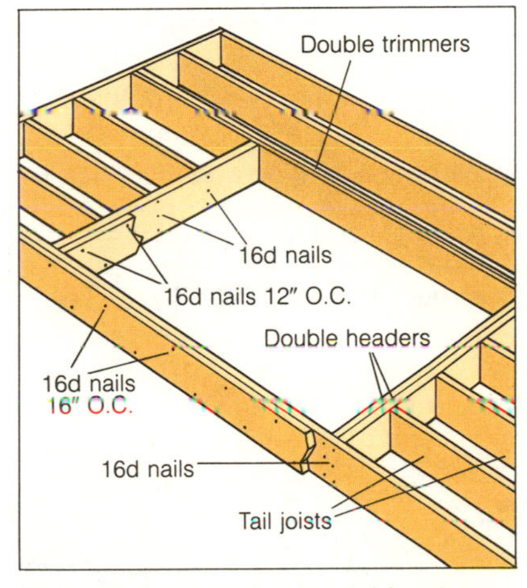

Floor openings require special framing.

Blocking or bridging? Solid blocking, 1 by 3 or 1 by 4 cross-bridging, or manufactured metal bridging will keep joists from rotating along their midspans. As a rule of thumb, install blocking or bridging at 8-foot intervals; if your span is less than 16 feet, split the distance.

Metal bridging is the simplest to install because it requires no nails; simply hammer the spiked ends into adjacent joist faces. To lay out a run of bridging, measure the desired distance in from both ends of the sill and snap a chalkline across the joists. (To make a working platform, span the joists with a long board near the line.) Attach the tops of the bridging, following the chalkline; leave a slight gap between adjacent pieces. Later, when the subfloor is down, hammer the bottoms into place from the crawlspace or basement.

You'll need solid wood blocking, cut from the same materials as the joists, between joists at the girder. Again, snap a center line. Toenail the blocks to the joists with three 8-penny nails at each end.

A JOIST SYSTEM OVERVIEW

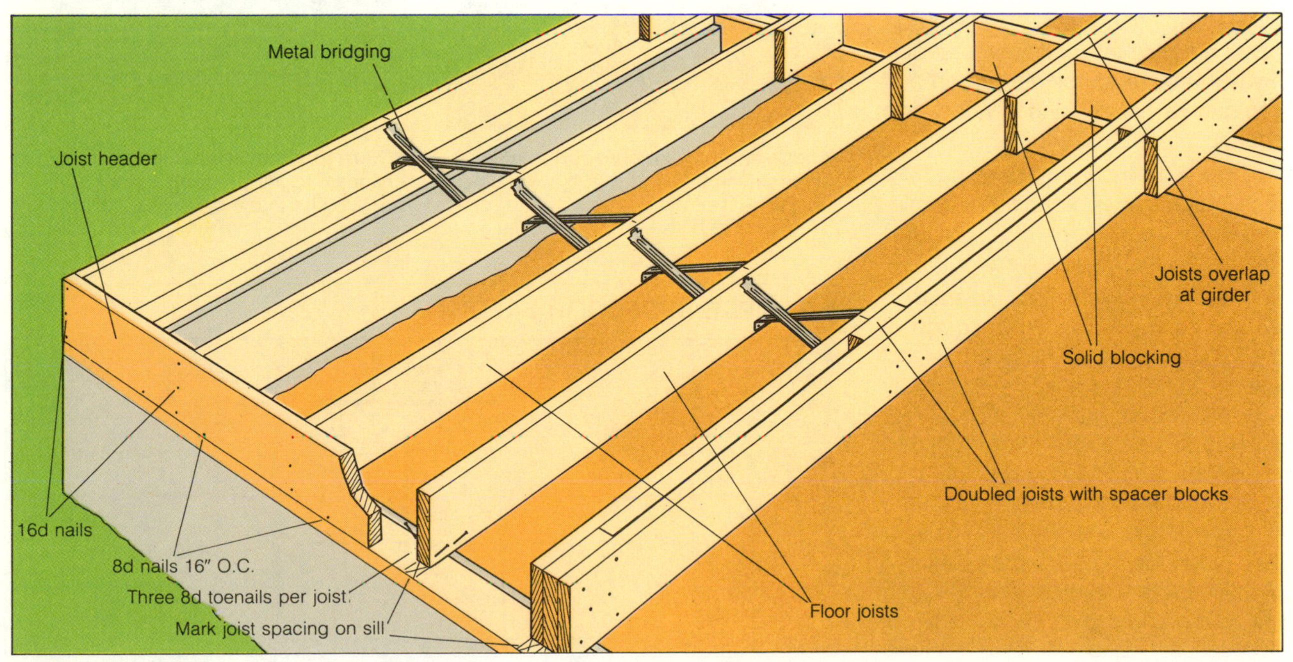

Nail floor joists both to joist headers and to sills at opposite sides of the building. Where joists overlap, nail them first to each other, then to the girder.

. . . Floor Framing

Laying the Subfloor

Completing the floor system is the floor deck. It is nailed directly to the tops of the joists. Though traditional solid board decking is still used (notably in post and beam designs), plywood has become the material of choice because of its structural rigidity and ease of installation.

Design notes. Most decks consist of two layers. The first layer, or subfloor, is 7⁄16-inch to 3⁄4-inch plywood—rated Sheathing, C-Plugged, or similar grade. A thinner, smoother, second layer, or underlayment, goes on just before the finish floor, creating a smooth surface.

Increasingly popular, though, are combination subfloor-underlayment panels. Ranging in thickness from 19⁄32 inch to 1 1⁄8 inches, these panels are available with tongue-and-groove edges that eliminate the need for edge support between joists. Additional underlayment is usually unnecessary.

Installing a plywood subfloor. Plywood panels are laid perpendicular to joists, with panel ends centered on joists. You don't need to support panel edges if the joints of the underlayment above will be staggered or if you plan to install 25⁄32-inch wood strip flooring.

For a 16-inch joist spacing, 7⁄16-inch sheathing is normally the minimum thickness; for 24-inch joist centers, 3⁄4-inch is required.

To install the subfloor, measure in 48 inches from the outside edge of the joist header at both ends and snap a chalkline across the joists. Lay the first row of plywood with edges flush to this line, and ends centered on joists. Panel ends in adjacent rows must be staggered: begin every second row with a half sheet.

When laying out the subfloor, simply tack the panels in place as you go, so that you can make adjustments if necessary. Later, snap a chalkline across joist centers and finish nailing. Use 6-penny common nails for panels up to 1⁄2 inch thick and 8-penny nails for thicker subfloors. Space nails 6 inches apart at panel ends and 10 inches apart at intermediate joists.

Installing subfloor-underlayment. Combination panels are laid out in the same manner as standard subfloors, but working with tongue-and-groove edges calls for a little more technique. Begin the first row by positioning the tongue to the outside, over the joist header. To fit the tongues of the second row into the grooves of the first, you'll need a sledge and a scrap 2 by 4 block. Use the sledge to rap the block up and down the groove edge, seating the tongue.

Choose 6-penny ring-shank nails for panels up to 3⁄4 inch thick, and 8-penny nails for thicker sheets. When tacking the sheets, don't nail the groove edge or you'll have trouble fitting in the next panel's tongue.

For an even sturdier deck, use both nails and an elastomeric construction adhesive. Gluing enables you to use fewer nails: space them 12 inches apart, both at panel ends and at intermediate supports.

LAYING PLYWOOD: TWO TIPS

Tongue-and-groove plywood panels can be driven together with a scrap block and sledge.

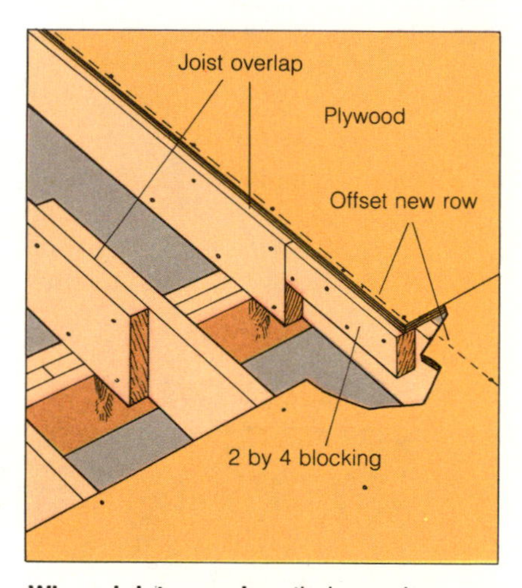

Where joists overlap, their spacing changes; to finish nailing a row, "scab on" 2 by 4s to the new joists.

A SAMPLE SUBFLOOR LAYOUT

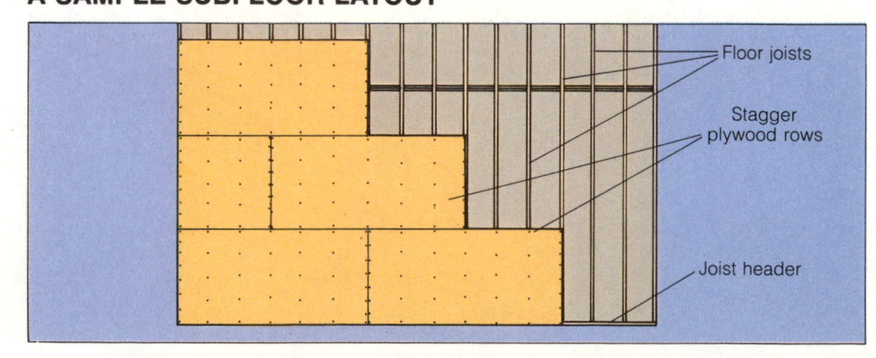

Lay plywood perpendicular to joists, with panel ends centered on joists. Stagger end joints by beginning alternate rows with a half sheet.

Wall Framing

With a solid subfloor beneath you, framing the walls for a new structure is a straightforward procedure that involves 1) assembling the wall components while they lie on the floor deck; 2) raising each wall section into position, checking it for plumb, and bracing it securely; and 3) fastening individual wall sections to the floor framing and to each other.

If you're starting from scratch, you'll assemble and raise the exterior side walls first, tie in the end walls, and then add interior partition walls. Remodeling is another matter. It often entails adding a new interior wall to redefine your space or redirect traffic flow.

New structure or old, here are techniques to help you assemble and raise walls.

Assembling New Walls

Framing for any wall includes a sole plate, evenly spaced wall studs, and a top plate. Horizontal fireblocks between studs are sometimes required. For walls with doorways or window openings, you'll need extra studs, as well as headers to span the opening.

Design notes. Traditionally, walls are built from 2 by 4 studs and plates, with studs placed on 16-inch centers. In recent years, the growing concern with energy-efficient structures has led to higher R-value recommendations (see page 53) than are possible with insulated 2 by 4 walls. Many builders respond by framing exterior walls with 2 by 6s, placing studs on 24-inch centers. Whichever design you choose, basic wall construction is the same.

Walls may be bearing or nonbearing. A bearing wall helps support the weight of the house; a nonbearing wall does not. All exterior walls running perpendicular to floor and ceiling joists are bearing. Normally, at least one main interior wall—situated over a girder or interior foundation wall—is also bearing.

How does this affect your planning? Design requirements for interior nonbearing walls—often called "partitions"—are less strict; partitions built from 2 by 4s on 24-inch centers are the norm. The framing for an opening—particularly the header—may be lighter as well.

The standard ceiling height for interior spaces is 8 feet. Because ceiling materials encroach on this height, you'll have trouble installing 4 by 8-foot gypsum wallboard or sheet paneling unless you frame the walls slightly higher (8 feet ¾ inch is standard). Subtracting the thickness of the sole plate and doubled top plates leaves a length of 7 feet 8¼ inches (92¼ inches) for the wall studs. Lumberyards frequently stock precut studs in this length.

Laying out the basic wall. First, cut both sole and top plates to length. If you need more than one piece for each, locate the joints at stud centers; offset joints between top and sole plates at least 4 feet.

Now lay the top plate against the sole plate on the deck, as shown below. Beginning at one end, measure in 1½ inches—the thickness of a stud—and draw a line across both plates with a pencil and combination square. If your studs are on 16-inch centers, start once more from that end; measure and draw lines at 15¼ and 16¾ inches.

From these lines, advance 16 inches at a time, drawing new lines, until you reach the far end of both plates. Each set of lines will outline the placement of a stud, with all studs evenly spaced 16 inches on center.

If you're installing studs on 24-inch centers, adjust the initial placement of lines to 23¼ and 24¾ inches. Then space lines 24 inches apart.

BEARING OR NONBEARING?

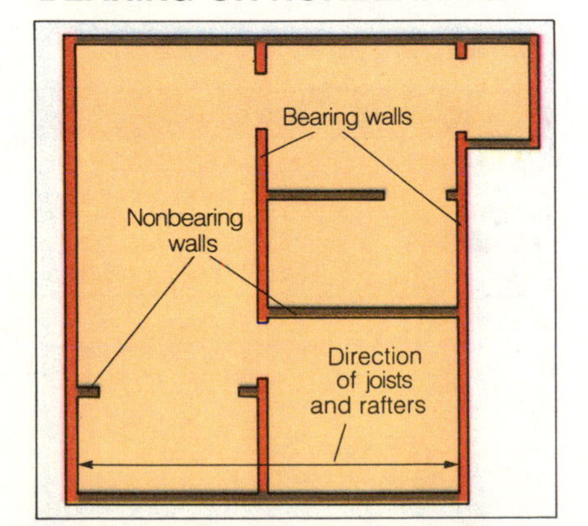

Walls are classified as either bearing or nonbearing; bearing walls support joists and rafters at their ends or at midspan.

MARKING STUD POSITIONS

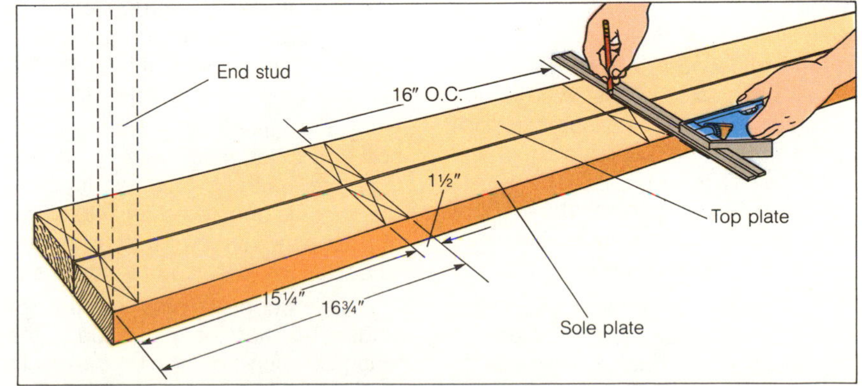

Lay sole and top plates together and mark stud positions on both at once. The spacings shown for studs on 16-inch centers. Continue marking lines at 16-inch intervals until you reach the end of both plates.

...Wall Framing

ASSEMBLING A BASIC WALL

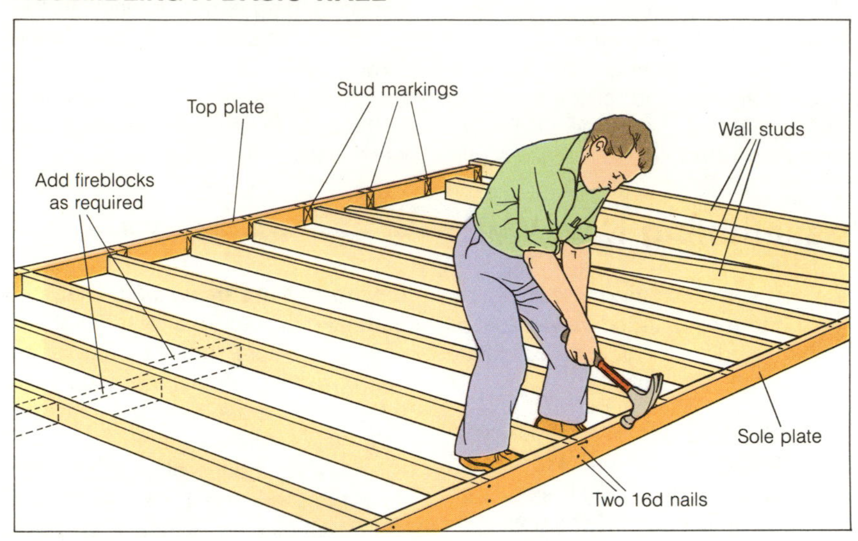

With plates marked and studs cut to length, lay the pieces out on the subfloor. Line up each stud and nail it to both plates. If fireblocks are required, add them next, staggering blocks slightly for easy nailing.

CORNER FRAMING

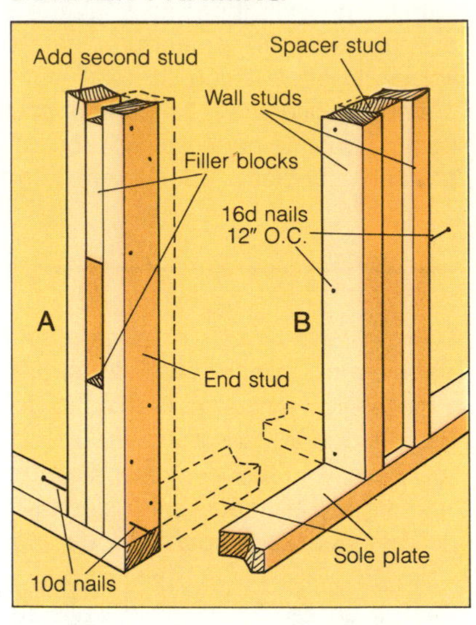

A wall intersection needs either a corner post (A) or, where a partition wall ties in, three studs as shown (B).

Assembling the pieces. Unless you're using precut studs, measure and cut the wall studs to exact length. Spread the plates apart on the deck and turn them on edge, stud markings toward the center. Place the studs between the lines and nail them through each plate with two 16-penny common nails or vinyl-coated sinkers.

If fireblocks are required, center them 4 feet above the bottom of the sole plate. Snap a chalkline across the studs and place alternate blocks slightly above and below the line. This allows you to angle nails through the studs from outside, rather than toenailing.

Framing corners. Where walls meet, you'll need extra studs. After a side (long exterior) wall is assembled, add extra studs at both ends; space these studs away from the end studs with filler blocks, as shown above right. Nail through both studs into each block with three 10-penny nails.

Wherever an interior wall meets the exterior walls, add three studs, as shown. Nail the outside studs to the spacer with 16-penny nails. You may find it simpler to add this assembly once the interior wall is raised.

Framing openings. If your new wall includes a door or window, mark the center line of the opening on the plates when laying out the wall studs. Be sure to check the manufacturer's "rough opening" dimensions—the exact height and width required—before framing the opening. If no specification is indicated, measure the unit and add an extra ⅜ inch all around for windows, or ½ inch at top and sides for doors. This additional space allows you to level and plumb the unit once installed.

- **To install rough door framing,** first measure off half the rough opening width in each direction from the center line, and draw a line; this marks the inside edge of each trimmer stud.

Now mark off both trimmer studs and king studs (see drawing on facing page) on the two plates. If possible, shift the opening slightly to utilize at least one standard stud as a king stud. Nail the king studs to the plate.

Next, cut and install the trimmer studs. Trimmer height equals the rough opening height *plus* the thickness of the finish floor and underlayment (see page 58). Now *subtract* the 1½-inch thickness of the sole plate and cut the trimmers to this length. Nail them to the king studs with 10-penny nails in a staggered pattern, as shown on page 71.

Bearing wall headers for 2 by 4 walls are typically composed of matching lengths of 2-by lumber turned on edge, with a ½-inch plywood spacer sandwiched between them. The exact depth of the required header depends on the width of your opening—and on your local building code. The chart below lists common header sizes.

MINIMUM HEADER SIZES

Opening Width	Header Size
Up to 4'0"	4 by 4 or two 2 by 4s on edge
4'0" to 6'0"	4 by 6 or two 2 by 6s on edge
6'0" to 8'0"	4 by 8 or two 2 by 8s on edge
8'0" to 10'0"	4 by 10 or two 2 by 10s on edge
10'0" to 12'0"	4 by 12 or two 2 by 12s on edge

Sizes are for 2 by 4 stud walls in single-story structures. If there's a second story above, choose the next larger header size.

To assemble a header, cut 2-bys and plywood to the length between king studs. Nail the pieces together with 16-penny nails spaced 16 inches apart along both top and bottom edges.

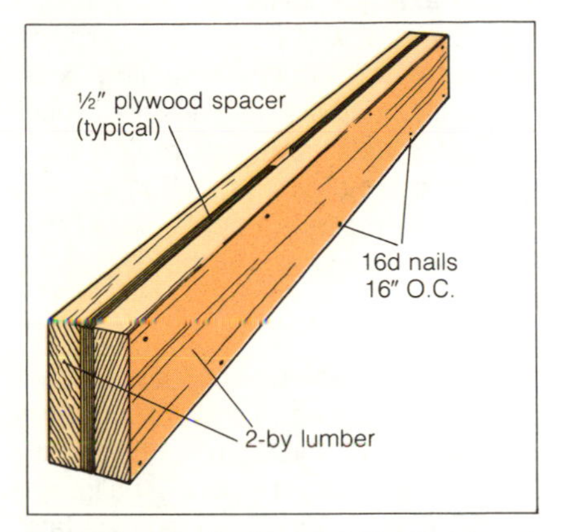

Build a header by "sandwiching" plywood between two 2-bys.

If your new walls are 2 by 6s, your header must be 5½ inches thick. The simplest way to accomplish this is to use one solid 6-by beam. A partition wall header may be a single 2 by 4 or 2 by 6 laid flat across the opening.

Place the header snug against the trimmers and nail it through the king studs with 16-penny nails. Measure and cut the cripple studs to length; if possible, use the same 16-inch spacing as the standard studs. Nail the cripples through the top plate with 16-penny nails; then toenail the bottoms to the header with 8-penny nails.

■ **Rough window framing** begins with placing the king studs, as described above. Nail the king studs to both plates.

Ideally, the tops of all doors and windows will be the same height—typically 6 feet, 8 inches. In practical terms, this means that the bottom edges of the headers should match. Measure the height from the top of the sole plate to the bottom of any door header you've installed. Mark this height on the window opening's king studs. Now, working down from the mark, subtract the height of the rough opening. This indicates the top of the rough sill. Next, subtract another 1½ inches for the sill's thickness and make another mark. The remaining distance to the top of the sole plate equals the height of the lower cripple studs.

After cutting the lower cripple studs to length, nail the outside pair to the king studs with 10-penny nails. Cut the sill and attach it to the king studs with 16-penny nails. Add the remaining lower cripples and, next, the trimmers on each side. Finally, install the header and upper cripple studs as described above.

Corner bracing. If your walls will be covered with plywood or diagonally laid solid lumber (see pages 82–87), you probably won't need to brace exterior walls. Any other materials require diagonal bracing—"let-in" 1 by 4s or steel straps—at all corners and every 25 feet along a wall. You'll need to notch (1 by 4s) or groove (ribbed straps) wall framing for the bracing; fasten both type to each stud or plate with two 8-penny nails.

ROUGH FRAMING FOR WALL OPENINGS

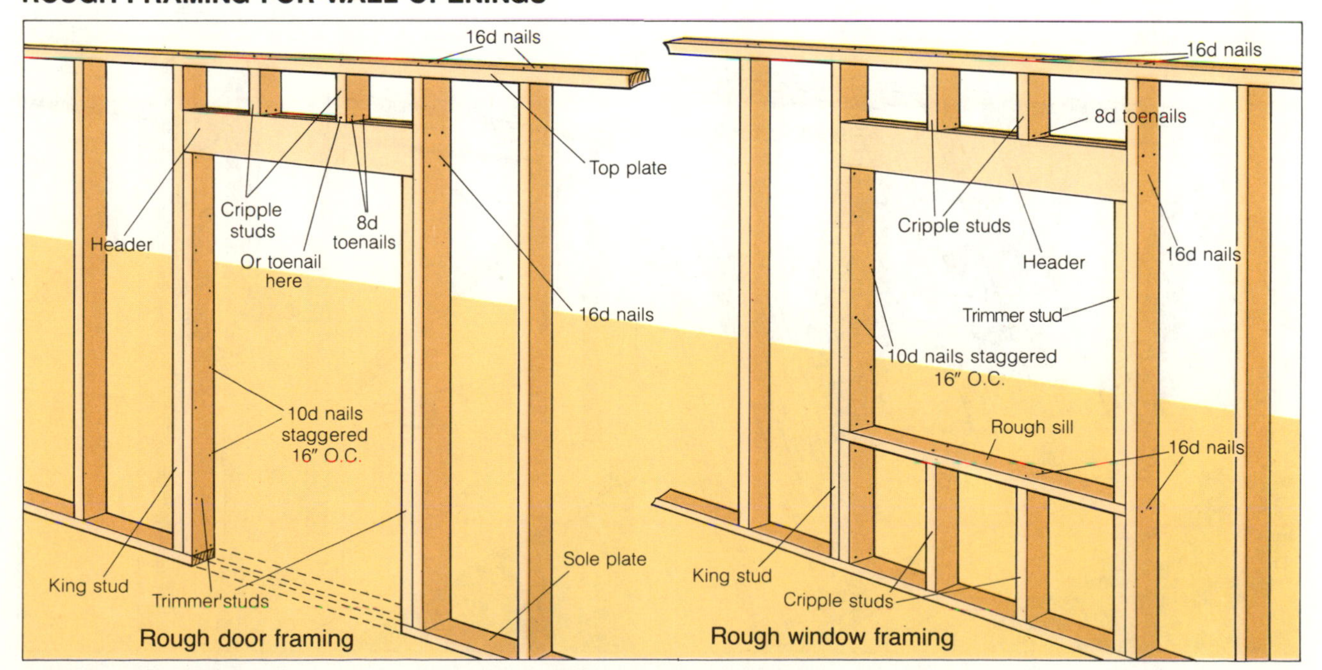

Doors and windows require special framing. A door opening (left) needs king studs, trimmers, header, and cripple studs. Window framing (right) includes a rough sill and cripples at top and bottom.

. . . Wall Framing

Raising the Walls

Once your first wall is assembled, only a few short steps remain until it's up and standing. Ideally, three willing workers should be on hand: one at each end to raise and hold the wall, and one in the middle to erect temporary bracing after the wall is up.

Up goes the wall. First snap a straight chalkline on the subfloor, indicating the sole plate's inside edge. Slide the wall along the floor deck until the sole plate lies near the edge of the deck. To prevent the plate from slipping off the platform, nail scrap blocks to the joist header or stringer, as shown below. Cut several 1 by 4 braces and 2 by 4 blocks.

To raise a wall, all the workers should grip it at the top plate and work their hands beneath the plate. Now everyone raises the plate in unison and "walks" down the wall until it's in the upright position. Next, tap the sole plate into line and tack it to the subfloor with a few duplex nails.

To brace the wall, tack three 1 by 4 braces to wall studs—one near each end and another in the middle. Nail the lower ends to the sides of 2 by 4 blocks. Eyeball the wall for plumb and then nail the 2 by 4s to the floor deck.

Check alignment. Using a carpenter's level, check the wall for plumb along both end studs on adjacent faces. Where the wall is out of plumb, loosen that brace, align the wall, and nail the brace once more. If an end stud is warped, "bridge" the warp with a straight board and two small blocks, as shown below. When both ends are plumb, adjust the middle.

Slab construction. When building atop a concrete slab, attach a wall's sole plate directly to preset anchor bolts—much like a standard mudsill (see page 64). Drill holes in the sole plate for each bolt; assemble the wall; raise and brace it as described above (nail braces to stakes *outside* the slab); then secure the washers and nuts on the bolts.

Fill in exterior walls. Before you anchor the first wall permanently, raise, brace, and plumb the remaining exterior walls. Install side walls first, then add end walls and any projections. Now check the diagonals from corner to corner across the deck. Adjust the walls as necessary until the diagonals match.

Anchor the walls. When everything is aligned, nail each wall through the sole plate into joists, joist header, or stringer, spacing 16-penny nails every 16 inches. *Don't* nail the sole plate within a doorway—that section will be cut away when the door is installed (see pages 94–96).

At corners, nail through the end walls into the corner posts in the side walls with 10-penny nails staggered every 12 inches.

RAISING, BRACING & PLUMBING A WALL

To raise a wall, lift it by the top plate; then, using the sole plate as a pivot, "walk" your hands down the wall until it's upright.

Brace the upright wall, then check plumb with a carpenter's level. A straight board and two small blocks help bridge warped studs for accurate readings.

Add interior walls. Before raising an interior wall, snap another chalkline on the floor deck to indicate the correct layout. Interior walls are raised, braced, and plumbed just like exterior walls. Where they intersect exterior walls, nail through the end studs into the exterior studs with 10-penny nails staggered every 12 inches.

Tie all the walls together. Are all the walls anchored? Then nail a second set of top plates—the "top cap"—onto the first, offsetting all joints below by at least 4 feet. At corners and intersections, overlap the joints below. Space 16-penny nails every 16 inches, and use two nails at joints and intersections. Leave all braces in place until the ceiling joists and rafters are in place.

Remodeling: Adding a Partition Wall

To separate one living space from another or to subdivide a room, you may need to build a new partition wall. Unlike walls for new structures, a partition for an existing house is usually built right in place. The basic steps are outlined below. To install framing for an opening, see pages 70–71.

Plotting the location. The new wall should be anchored securely to the floor, to the ceiling joists, and to wall framing on at least one side.

To locate existing wall studs, try knocking with your fist along the wall until the sound changes from hollow to solid. If you have wallboard, you can use an inexpensive stud finder, though the nails that hold wallboard to the studs are often visible on close inspection. Find studs behind plaster walls by driving a small test nail just above the baseboard.

To locate ceiling joists, use the same methods or, from the attic or crawlspace above, drive small nails down through the ceiling on both sides of a joist to serve as reference points. Adjacent joists and studs will be evenly spaced, usually 16 or 24 inches away from those you've located.

A wall running perpendicular to the joists will be the easiest to attach. If wall and joists will run parallel, though, try to center the wall under a single joist; otherwise you'll need to install nailing blocks every 2 feet between two parallel joists (see illustration at right). If the side of the new wall falls between existing studs, you'll need to remove wall materials and install additional nailing blocks.

On the ceiling, mark both ends of the center line of the new wall. Measure 1¾ inches (half the width of a 2 by 4 top plate) on both sides of each mark; snap parallel lines between corresponding marks with a chalkline.

Positioning the sole plate. Hang a plumb bob from each end of the lines you just marked and mark these new points on the floor. Snap two more chalklines to connect the floor points.

Cut both sole plate and top plate to the desired length. Lay the sole plate between the lines on the floor and nail it in place with 10-penny nails spaced every 16 inches. If you're planning a doorway, don't nail through that section of the plate; it will be cut out later. (With masonry floors, use a masonry bit to drill holes through the sole plate and into the floor every 2 or 3 feet. Then insert expanding anchors.)

Marking stud positions. Lay the top plate against the sole plate, as shown on page 69, and mark the stud positions. If local codes permit, use 24-inch spacing (you'll save lumber) and adjust the initial placement of lines to 23¼ and 24¾ inches.

Fastening the top plate. With two helpers, lift the top plate into position between the lines marked on the ceiling; nail it to perpendicular joists, to one parallel joist, or to nailing blocks, as shown at right.

Attaching the studs. Measure and cut the studs to exact length. Attach one end stud (or both) to existing studs or to nailing blocks between studs. Lift the remaining studs into place one at a time; line them up on the marks and check plumb with a carpenter's level. Toenail the studs to both top plate and sole plate with 8-penny nails. Some building codes require horizontal fireblocks between studs. If permitted, position blocks to provide an extra nailing surface for wall materials.

TOP PLATE OPTIONS

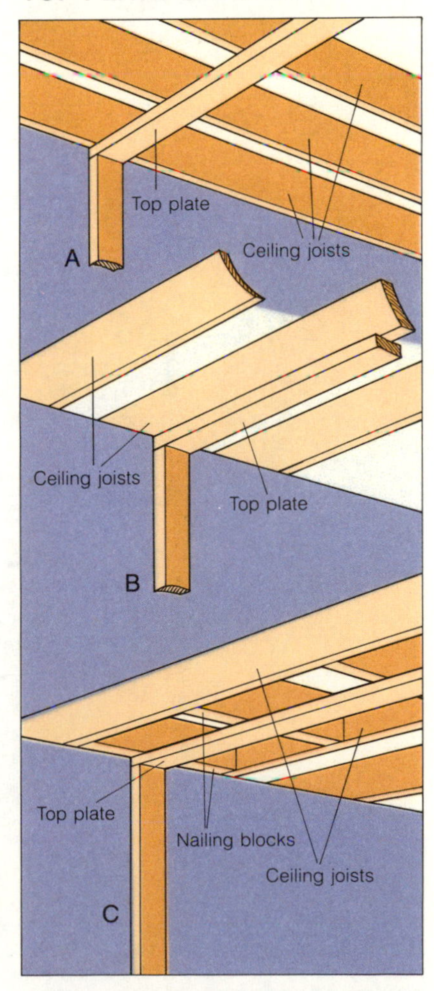

To anchor a top plate, nail it to perpendicular joists (A), to the bottom of one parallel joist (B), or to nailing blocks (C).

READING PLANS: AN INTRODUCTION

Though few homeowners may have the knowledge or skills to draw up a really accurate set of blueprints—especially for a large project—understanding blueprints is essential for anyone planning to take on much of the work. Here's a quick course on reading house plans.

For any complex project, you'll most likely be referring to a number of scale drawings, including some or all of the following:

Plot plan. This overview shows the relationship of the building or improvement to property lines, topography, easements, and the like.

Floor plan. This classic bird's-eye view of the structure's layout shows dimensions and positions of walls, windows, doors, and other features. If the space is uncomplicated, the plan may include plumbing and electrical layouts as well.

Elevations. Elevations are straight-on views of single interior or exterior walls that help explain the floor plan. They often specify dimensions and materials, in addition to pointing out special design features.

Foundation and framing plans. These specialized drawings, shown from the same vantage point as the floor plan, feature foundation layout or framing members—floor, wall, or roof. They allow you to quickly visualize structural details and estimate your needs.

Sections and details. Whenever standard views or scales can't show fine points, "sections" or "details" are added. Sections are planes sliced vertically or horizontally through the building to indicate, for example, windows or foundation footings. Large-scale details—normally side views—show trim, fastenings, and dimensions in great detail.

THREE VIEWS OF ONE ROOM ADDITION

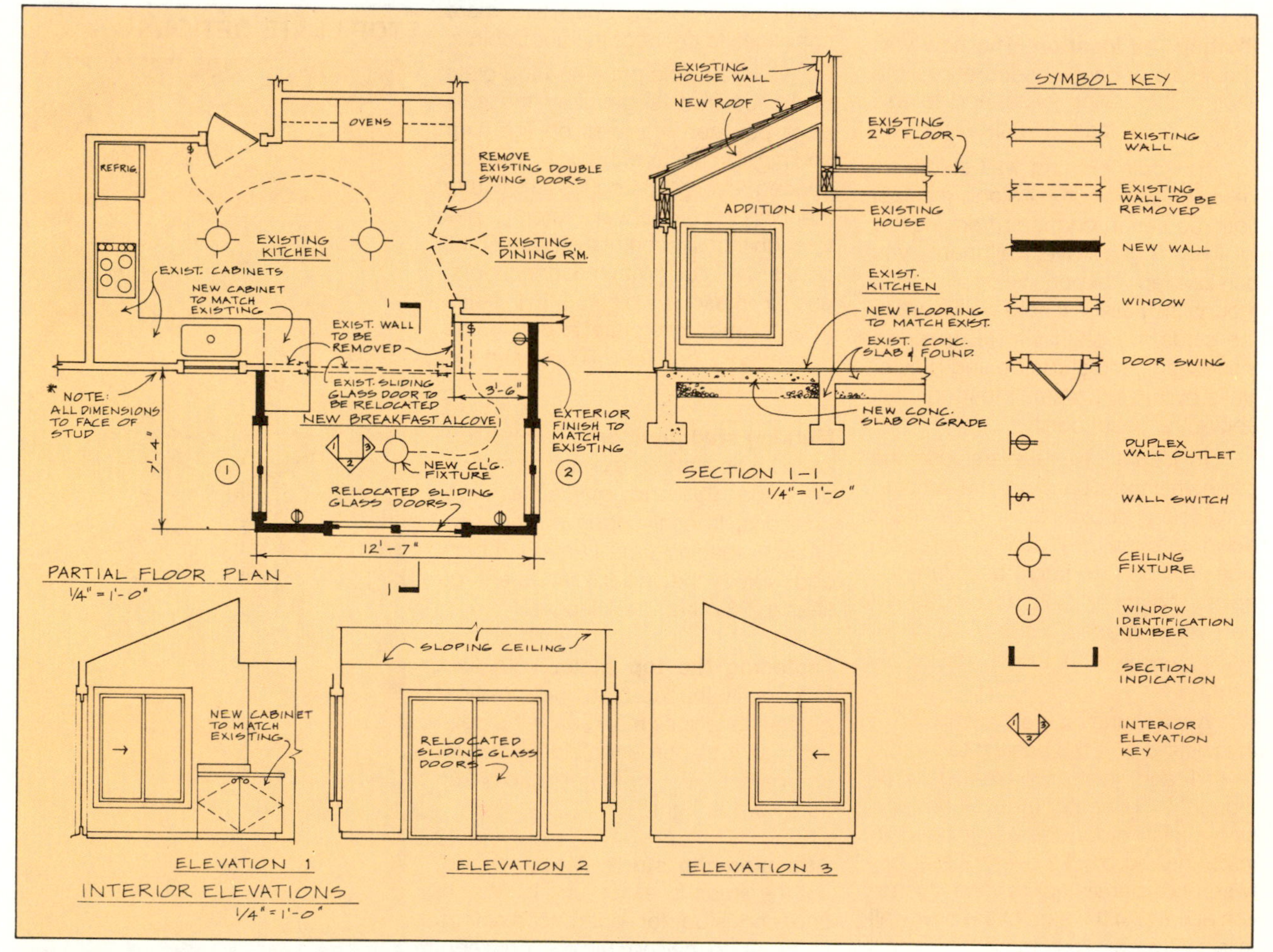

Floor plan, elevation, and section views combine to show the "full picture" of a proposed addition. At right is a key to the standard architectural symbols used in the drawings.

Ceiling & Roof Framing

Roof framing—the number one nemesis of many carpenters—needn't be mysterious; it's based largely on a few standard principles, careful layout, and hard work. In this section we'll introduce those principles and walk you through the construction of a straightforward gable roof.

The sequence includes installing ceiling joists; measuring, cutting, and assembling rafters; and finishing up the job with gable studs, barge rafters, and collar beams. You can apply these basics to whatever project you have in mind: greenhouse, garage, or lakeside cabin.

Before beginning work, check the walls for square once more by measuring diagonals from corner to corner across the top plates.

Safety is the major concern when framing any roof. Always take time to plan the proper operating sequence *before you start.*

Installing Ceiling Joists

Ceiling joists serve two main functions: they support the load of the ceiling materials below, and they help to permanently brace exterior walls against the thrust of the rafters above. Think of ceiling joists as lightweight versions of floor joists (see pages 66–67). Like floor joists, they normally span the structure's short dimension (in this case, resting on opposite exterior walls or atop an interior bearing wall at one end).

Design notes. Ceiling joists are typically 2 by 4s, 2 by 6s, or 2 by 8s spaced 16 or 24 inches apart on center (O.C.). The size you'll need depends on a combination of factors: lumber species and grade, the span, the spacing between joists, the type of ceiling below, and how much traffic or storage you anticipate in the attic or crawlspace above. You'll find that building codes contain a number of specialized span charts, allowing you to choose the exact size for your combined requirements.

Laying out the joists. The positions of ceiling joists are laid out along the top plates with a tape measure, combination square, and pencil (see drawing, page 69). It's safest to stand on a firm ladder while you mark the spacings. Lay the rafters out at the same time (see page 78). If their spacings are the same, place rafters and joists next to one another. This way, when the rafters are installed, they can be nailed to the joists. If the spacings are different—24 and 16 inches O.C.—space them so they'll meet every 48 inches.

Though it's not required, many carpenters position a joist just inside the top plate on each end wall, providing a solid nailing base for ceiling materials at the edge.

Because rafter spacings are identical on opposite plates, it's awkward to overlap ceiling joists where they cross an interior bearing wall or beam—they won't meet the rafters. Instead, plan to butt long spans together, as shown below.

Installing the joists. First cut all the joists to length; then pull them up into position and nail them down. This takes two carpenters working from ladders at opposite ends. Toenail the joists to each plate with three 8-penny nails. If the joists butt together at the center, nail each to the plate or beam and then tie them together with blocking, as shown below.

When a partition wall's top plate falls between parallel joists, install nailing blocks every 2 feet (see drawing, page 73), and toenail through the blocks into the plate.

If your plans call for a ceiling opening, frame it in the same manner as you would a floor opening (see page 67 for details).

LAYING OUT THE JOISTS

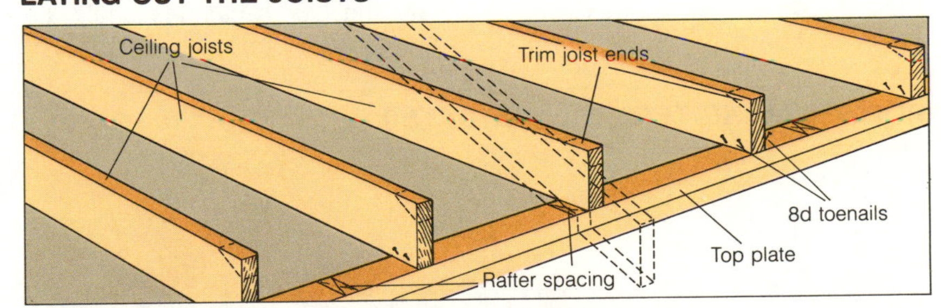

Lay out ceiling joists and rafters, then toenail joists to the top plate. Later, when rafters are installed, trim joist ends flush with the rafter slope.

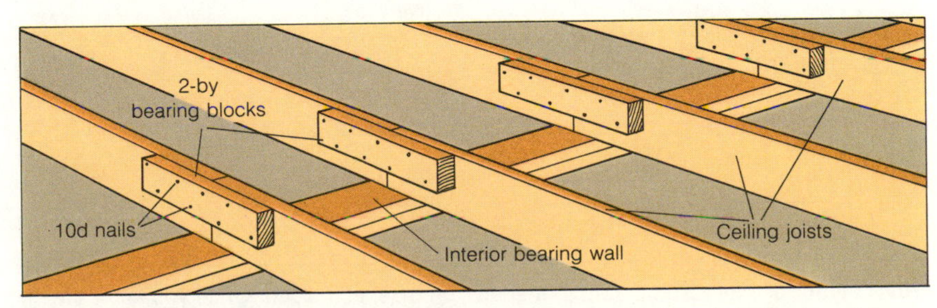

When joists cross a bearing wall, butt them together. Toenail each to the top plate and cover the joint with 2-by bearing blocks.

...Ceiling & Roof Framing

Roof Framing: An Introduction

A walk around your neighborhood will quickly reveal how many different roof configurations there are. Most are variations on the basic *gable* roof, which consists of evenly spaced pairs of *common* rafters running from the top plates to a central ridgeboard at the peak. Once you've grasped the principles of the gable roof, more complex designs should become clear. Here are the ABCs of laying out a gable roof.

Design notes. Typical rafter sizes are 2 by 4, 2 by 6, and 2 by 8, installed on either 16 or 24-inch centers. As with joists, the correct rafter size depends on span, spacing, and the load to be carried. In addition, you'll need to consider the roof's slope and whether or not you're in snow country. The local building department will have all the variables worked out for your area.

At the peak, rafter pairs butt against a central ridgeboard—either 1-by or 2-by lumber. Choose a ridgeboard one width larger than the rafters—for example, if you have 2 by 6 rafters, use an 8-inch wide (nominal size) ridgeboard.

The roof framer's glossary. Terms you'll need to understand include span, run, and rise. *Span* is the distance between the outside edges of the exterior top plates. *Run* equals half the span, or the "midspan"—the point directly below the roof peak. *Rise* is the vertical height of the peak at midspan, measured from the top plate.

Note, in the drawing below, that the outline of the run, the rise, and a rafter's top edge form a right triangle. Then remember the Pythagorean theorem for right triangles: $A^2 + B^2 = C^2$. What does this have to do with hammers and nails? If the run and rise—A and B in the drawing—are known, then the length of C—the rafter—can be quickly calculated.

The *slope* (pitch) of a roof is usually simplified in terms of *unit rise* and *unit run*. Unit run is always 12 inches; unit rise equals the slope in those 12 inches. You'll hear a roof described as—for example—"5 in 12." This means that the roof rises 5 inches vertically for every foot of run.

A common rafter requires three cuts: the *plumb cut* (the angle that meets the ridgeboard), the *bird's mouth* (the notch that fits over the top plate), and the *tail cut* (for the overhang that forms the eave).

When you lay out a rafter, your job is to measure, mark, and cut one rafter perfectly, and then use that rafter as a pattern for cutting the rest. Two layout techniques are commonly used: the "stepping off" method and the "rafter table" method. Both depend on the carpenter's old standby, the framing square.

Stepping off a rafter. Pick a straight piece of lumber for your pattern; if there's a crown (see page 38), place the crown side away from you (the rafters will be installed crown side up). Now, align the framing square so that the figure for the unit run—12 inches—and the figure representing the unit rise (5 in our example) on the outside of the body and tongue, respectively, meet the rafter's edge. To mark the plumb cut, trace a pencil line along the tongue. Then draw a line along the body.

Move the framing square down the rafter until the 5 on the tongue intersects the line you drew along the body at the rafter edge; line the body

ROOF-FRAMING TERMS ILLUSTRATED

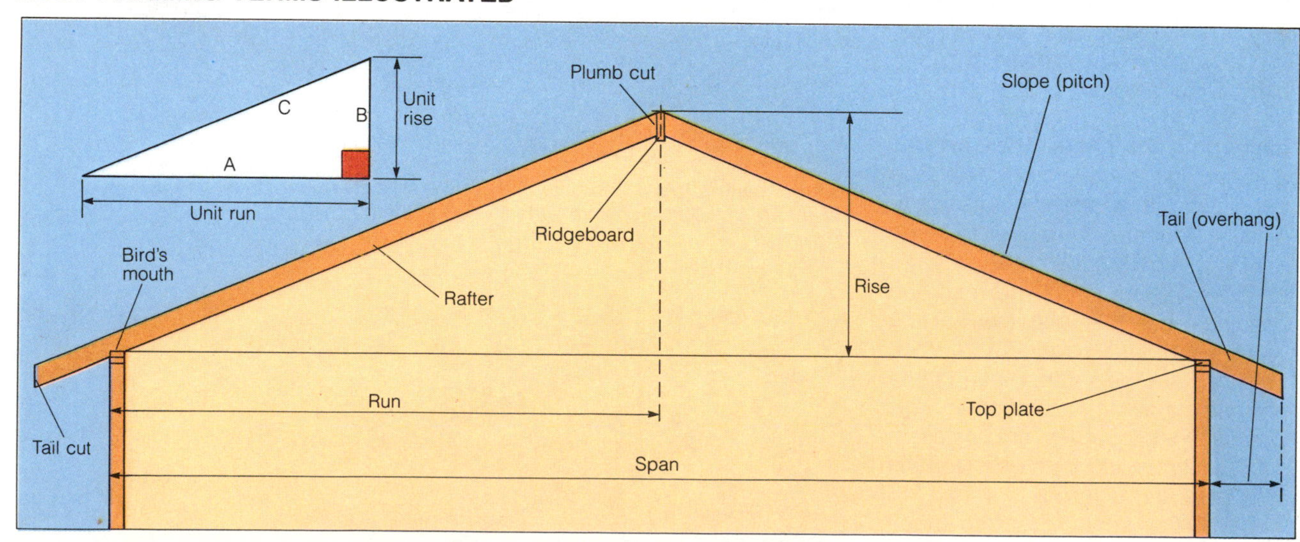

Roof layout is expressed in terms of span, run, rise, and slope. Once you know the rise and run, it's easy to find both slope and rafter length. When laying out rafters, you'll be dealing with plumb cuts, bird's mouths, and tail cuts.

STEPPING OFF A RAFTER

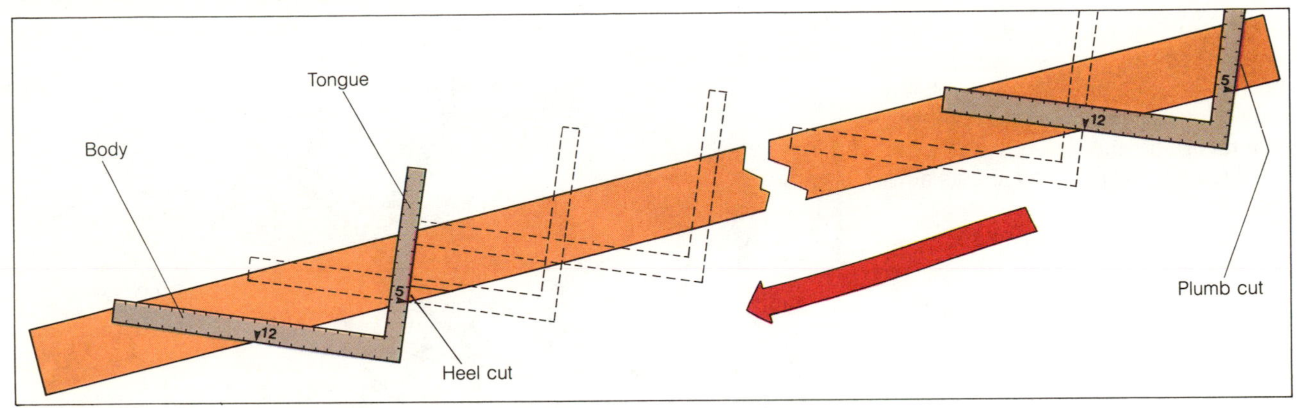

To step off a common rafter, align the framing square so the unit run (body) and unit rise (tongue) figures meet the rafter edge. Draw a line along the body, then slide the square down the rafter until the rise figure intersects the line at the rafter edge. Step off one increment for each foot of rafter run.

up with the 12 figure again and draw another line along the body.

Continue stepping off increments in the same way. If you're measuring a run of 10 feet, step off 10 times altogether. Then align the square once more and draw a new line along the tongue; this line indicates the *heel cut* of the bird's mouth.

To lay out the horizontal *seat cut* of the bird's mouth, measure up 1½ inches from the bottom end of the heel line. Slide the square back toward the top of the rafter, keeping the figures lined up, until the edge of the body intersects the 1½-inch mark. Draw the seat cut line along the body.

To mark the seat cut of a bird's mouth, align the square's body with the 1½-inch mark on the heel cut line.

What if your actual run is an odd increment like 10 feet 8 inches? On your tenth step with the square, extend the line along the body, then make a mark at 20 inches—or 8 extra inches. Now slide the square down the rafter until the tongue aligns with this mark, and draw the heel cut.

Are you planning an overhang? Step this off from the heel cut. The rafter tail may be cut plumb, level, or perpendicular to the rafter.

The total rafter distance you've laid out represents the theoretical length—as if there were no ridgeboard at the peak. You must now go back and subtract one-half the thickness of the ridgeboard—⅜ inch for a 1-by ridgeboard, ¾ inch for a 2-by—and draw a parallel line inside the original plumb cut line. (Measure this distance perpendicular to the first line—*not* along the rafter's edge.) This second line marks the actual length of the rafter.

Using the rafter tables. You can also figure a rafter's length by consulting the tables on your framing square. Below each inch mark on the body's face side, you'll find a column of six or seven figures. The inch marks on the body serve as an index to unit rise; in our example, we'd check the column under the 5-inch figure. The top set of numbers represents the "length common rafters per foot run." Thus, for every 12 inches of unit run and 5 inches of rise, the rafter length will be 13 inches. Multiply this figure by the 10 feet of run in our example, and you have 130 inches—or 10 feet 10 inches. (NOTE: The charts on your square may differ from those discussed above; consult the instruction booklet that comes with the square.)

To lay out the rafter, first draw the plumb cut at the top by aligning the square as described under "Stepping off a rafter" (preceding). Now measure the actual distance (10 feet 10 inches) along the rafter edge to the heel cut. Again, lay out the heel and seat cuts as described earlier. Finally, add on the tail's length. Be sure to subtract half the thickness of the ridgeboard from the theoretical length you've laid out.

Cutting the rafter. Although plumb and tail cuts are straightforward, a bird's mouth requires special care. The following method produces neat results. Beginning with a circular saw, cut along the heel and seat cut lines *only* to the point where they intersect. Finish the corner with a crosscut saw held upright. If a rough edge remains, smooth it with a sharp chisel.

...Ceiling & Roof Framing

Assembling the Gable Roof

After your first rafter is laid out and cut, make a duplicate so you'll have a pair. To check your work, assemble a couple of willing helpers and take the rafters up on the ceiling joists. For solid footing, lay plywood sheets atop the joists.

Attach a scrap block of the same material you'll use for the ridgeboard to one rafter's plumb cut. Then raise both rafters into position. Are they snug against the top plate at both ends and flush with the ridgeboard scrap? If so, take them down and cut the remaining rafters.

Next, measure and cut the ridgeboard to length, transferring the rafter spacings to its sides. If your structure is too long for a single ridgeboard, plan the joint so that it falls between a rafter pair. If you're planning a gable overhang (see "Framing a gable overhang," below), the ridgeboard must extend the appropriate distance beyond the end rafters.

RAISING THE ROOF

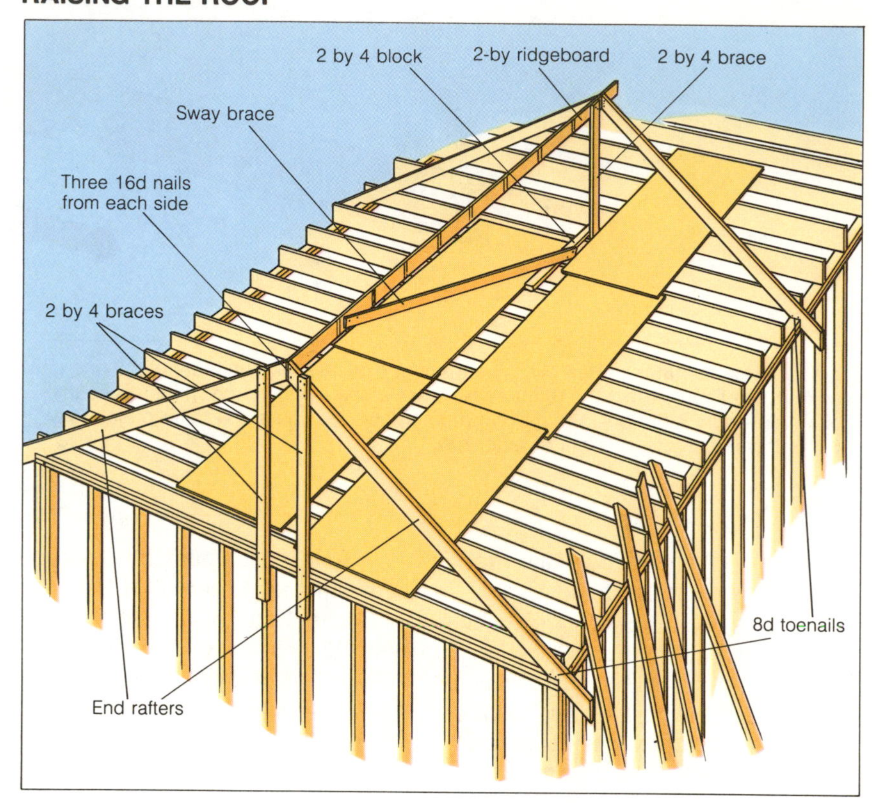

Assembling rafters calls for a deliberate work sequence. Here, the end rafters and the first pair back from the ridgeboard's end have been erected and carefully braced. The next steps are to fill in the rafter pairs and add the second ridgeboard section.

Raising the roof. To begin the roof-raising, first nail two upright 2 by 4s—one for each of the end rafters—flush against the wall's top plate, as shown above right. Prop each rafter against the top plate near its intended mark; when you need a rafter, just pull it up into position.

Now align the first end rafter flush with the end wall's plate. It takes three people to raise a large rafter. The first person lines the rafter up with the end of the plate; the second raises it into position. Be sure the bird's mouth is snug on the top plate. Then toenail the rafter to the plate with three 8-penny nails or use a metal framing connector. The third person can now tack it to the 2 by 4 brace.

While a helper supports the far end of the ridgeboard, raise it into position, align it with the top of the first rafter, and nail it to the rafter with three 8-penny (for a 1-by ridgeboard) or 16-penny (for a 2-by) nails.

If the ridgeboard is a single piece, repeat the procedure at the other gable end. If it's in two pieces, brace it at the far end, as shown, so the ridge is level; then attach one rafter at the first spacing back from the end.

Now go back and install the matching rafters, toenailing them to the ridgeboard from the opposite side. Then check your work: the end rafters should be flush with the end of the wall, and the ridgeboard must be both level and centered over the mid-span. The surest way to check is with a plumb bob.

When everything checks out, brace the assembly with a *sway brace* running diagonally from the ridgeboard to a 2 by 4 block nailed across the joists. Add the remaining rafters to the run in pairs, fastening each to the top plate and then to the ridgeboard.

If you need to add a second ridgeboard, proceed as before from the other direction. Be sure the boards' junction is covered by two rafters. Where rafters meet ceiling joists, tie them together with three 10-penny nails; then cut the joist ends to match the rafter slope

Filling in the gable ends. With the rafters in place, it's time to add gable studs between the end rafters and top plate.

Starting at the point directly below the peak, center a stud spacing on this spot. Now move toward both ends of the plate, laying out stud spacings on 16-inch centers.

The first stud fits securely between ridgeboard and top plate. To lay out the remaining studs, position one on its mark, flush against the back of the end rafter, and check it for plumb. Trace the rafter angle onto the edge of the stud. Then move to an adjacent stud and mark again. The difference in length between these studs is known as the *common difference;* it will be consistent between each adjacent stud in the row.

Cut the studs in pairs at the correct angle (a T-bevel will help you transfer the angle). Toenail each stud to the plate and rafter with 10-penny nails.

If your plans call for a gable end vent, frame the opening with horizontal 2 by 4s, bridging the studs as shown on page 56.

Framing a gable overhang. To form a roof overhang—or "rake"—at the gable end, you'll need to add outriggers and a pair of barge rafters to the ridgeboard extending past the end walls. Outriggers are 2 by 4s laid flat and positioned perpendicular to the rafters. Spaced every 4 feet down from the ridgeboard, they normally begin at the first rafters in from the end rafters; the end rafters are notched so the outriggers sit flush with the roof plane.

Nail outriggers to the first pair of rafters and then into the end rafters at the notches, using 16-penny nails. Leave them slightly longer than the ridgeboard; snap a chalkline 1½ inches in from the end of the ridgeboard and cut the outriggers off evenly.

Position the barge rafters as shown below, nailing them to the ridgeboard and into the ends of the outriggers. NOTE: Be especially cautious when installing a gable overhang, for this is the most exposed area on the entire roof.

Installing a collar beam. Roof structures with long spans—or without ceiling joists—are often tied together with collar beams bridging every third pair of rafters.

To install a collar beam, measure down one-third to one-half the length of each matching rafter. Hold a 1 by 6 board against the marks and trace the rafter angles on the back. Cut the collar beam on these lines and nail it to each rafter with 10-penny nails.

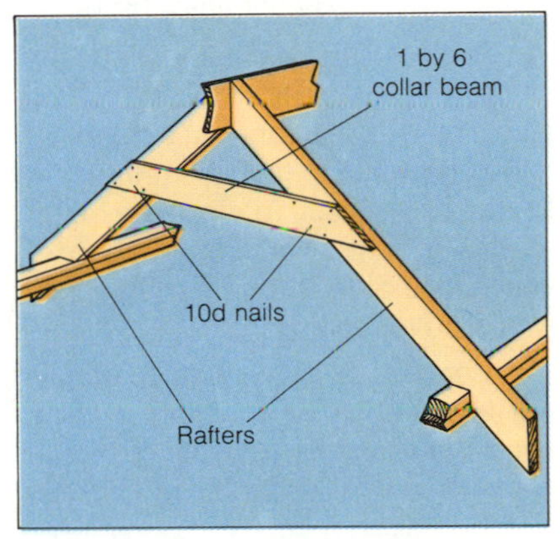

Collar beams, typically cut from 1 by 6 lumber, help reinforce rafter pairs.

Installing a purlin. When rafters exceed the maximum allowable span, they must be supported at midspan by horizontal *purlins* and braces, as shown below. As a rule, choose a purlin the same size as the rafters. Braces must sit atop a bearing wall and may be nailed directly to the plate or to ceiling joists.

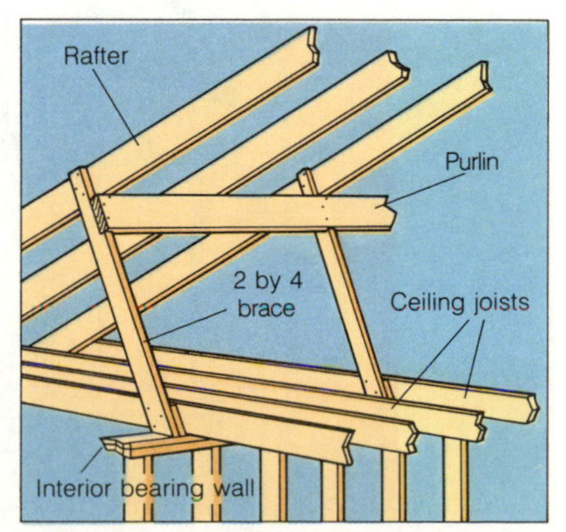

Purlins and braces support rafters that exceed their maximum allowable span. The purlins should be the same size as the rafters.

FILLING IN A GABLE END

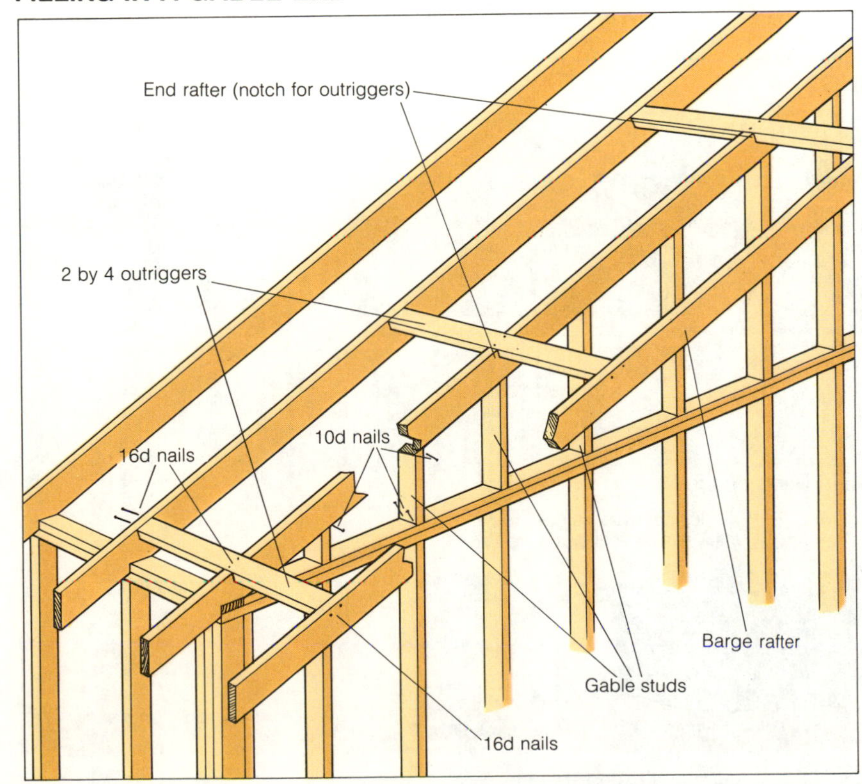

Gable end studs fill the gap between end rafters and top plate; figure the angle, cut them in pairs, and toenail them in place. Outriggers and barge rafters support a gable overhang.

INSULATING THE SHELL

An energy-efficient structure, new or old, calls for proper insulation. And the best time to insulate is during new construction or remodeling, when the spaces between joists, studs, and rafters are exposed.

Instructions are given here for mineral-fiber blankets or batts, the simplest type to install in exposed areas. If you need to add insulation to an existing structure in which spaces are concealed, see the *Sunset* book *Insulation & Weatherstripping* or consult an insulation contractor for help.

Tools of the insulation trade include a sharp utility knife for cutting blankets or batts and a lightweight hand stapler for fastening them to studs or joists. To insulate dark areas—in attics or under floors—you'll need at least one portable light with a 50-foot cord.

Because all insulation materials are either dusty or irritating to skin, eyes, nose, and lungs, wear gloves, a painter's mask, and plastic goggles when installing them.

Begin with the attic

Not only is an unfinished attic one of the easiest areas to insulate, but it yields the greatest energy savings relative to its cost.

Unless your attic will be used as living space, you need to insulate only between ceiling joists. The ceiling between the joists won't support you, so you'll need to lay a couple of boards over the joists to serve as temporary flooring.

If your blankets or batts have an attached vapor barrier (see page 53), lay them between the joists with the vapor barrier facing down. When you're working with a separate vapor barrier, fasten it first to the sides of the joists and then lay batts or blankets on top. Be sure the insulation extends over the top plates below, but don't cover eave vents or block their flow.

Also, be careful not to cover anything that can produce heat, such as electric fans and recessed light fixtures. Peel back the flammable vapor barrier 3 inches from chimneys, stovepipes, and flues. When possible, slip your batts underneath electrical wiring.

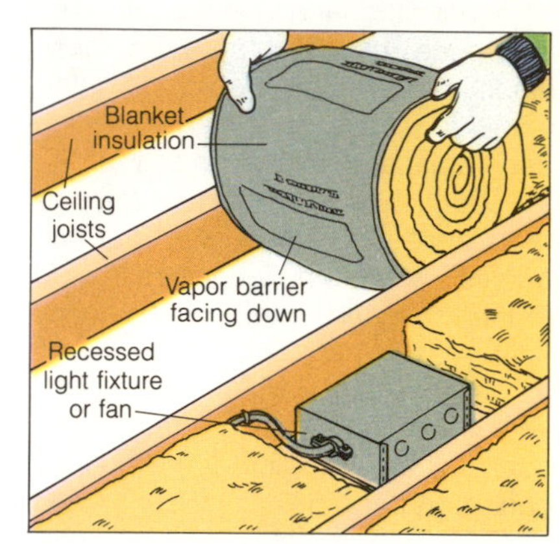

Unfinished attics call for blankets or batts between ceiling joists. Be sure not to cover heat-producing fixtures or fans.

If your attic is to be finished, plan to insulate between rafters, collar beams, short "knee wall" studs, and gable studs instead of ceiling joists. Again, the vapor barrier faces in, to-

WHERE DOES INSULATION BELONG?

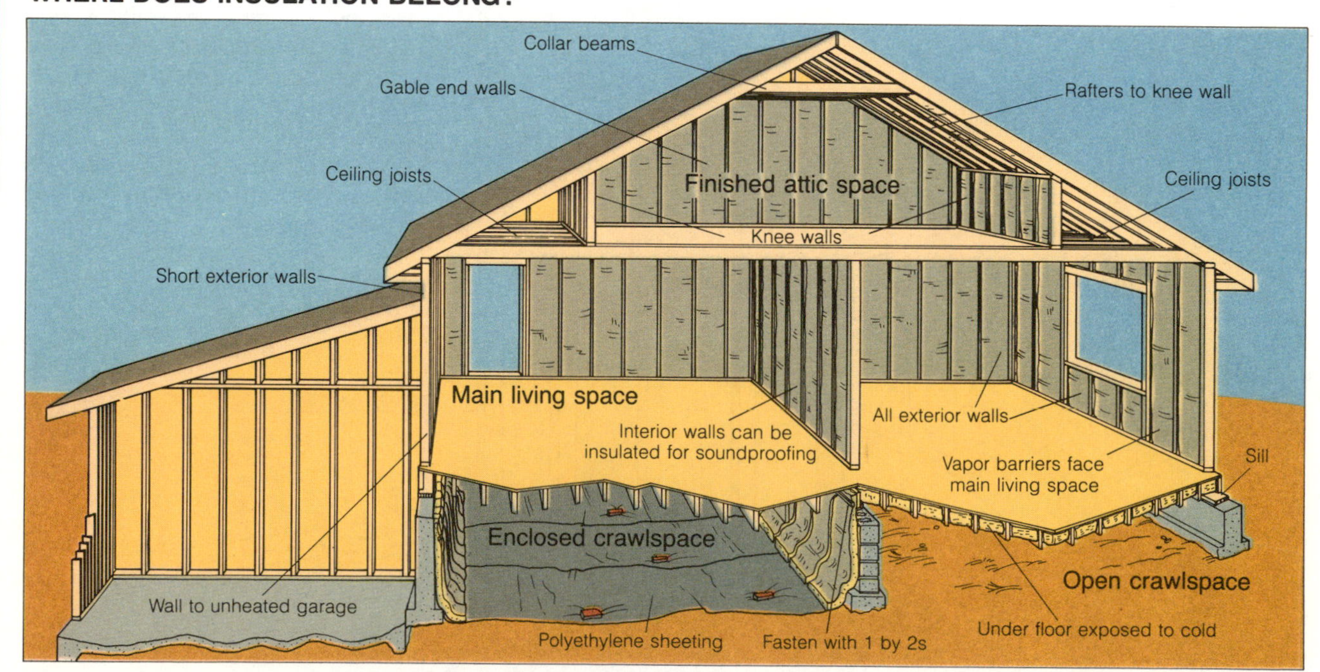

Install insulation inside all exterior stud walls, in attics, under floors exposed to the outside, and on enclosed crawlspace or basement foundation walls.

ward the attic. You can staple batts or blankets directly to the edges of rafters and studs, but be sure not to block any vent openings at the eaves. If you need to add a vapor barrier, staple it in place after insulation is installed.

Walls are next

Because walls make up such a large portion of your house exterior, they are second only to the roof in terms of losing or gaining heat.

Precut 4-foot batts are simplest to handle when you're insulating a standard 8-foot wall with fireblocks at 4-foot heights, but blankets can be easily cut to length. It's best to choose the type with an attached vapor barrier.

You can cut blankets or batts to length either on a subfloor or after they are partially placed in the wall, using the top plate, sole plate, or fireblock to back the cut. Size the pieces slightly long to ensure a tight fit.

Exposed wall framing is easy to insulate: staple vapor barrier flanges to studs and tuck in loose ends.

The attached vapor barrier should always face the side that's warm in winter, so it's simplest to place—and staple—insulation from the inside. If you must work from outside, fasten batts in place with one of the methods described below for floors.

Peel back flammable vapor barriers 3 inches from flues, chimneys, electric fans, and other heat-producing equipment. Stuff insulation scraps into cracks and small spaces between rough framing and the jambs of windows and doors. Also stuff spaces behind electrical conduit, outlet and switch boxes, and other obstructions.

Floors and basement walls

If heating costs are high where you live, don't neglect the areas under floors. Insulating foundation walls may also be advantageous in an enclosed crawlspace or heated basement that projects well above grade level. Because these areas are usually accessible, you can weatherproof them without much difficulty.

In new construction, you should insulate between the floor joists over an unheated area before nailing down the subfloor. This way you can staple insulation with an attached vapor barrier (which must face upward) to the joists with little effort.

To insulate existing floors, you'll have to work from below. Precut 4-foot batts are easiest to handle in cramped quarters. Otherwise, cut blankets to the exact length needed for each space. If you can find them, "reverse-flange" batts are excellent, since you can staple them from below.

There are several methods for holding standard blankets and batts in place. The most common methods include lacing 18-gauge baling wire back and forth between nails hammered into joists, and simply stapling chicken wire to the joists.

Contractors use stiff wire braces, cut 1 inch longer than the space between joists. You can cut your own from 13-gauge wire or even from wire hangers. To install insulation in this manner, hold the blanket or batt in place between joists and bow a wire brace gently up against the fibers at one end. Place an additional wire every 1½ feet. Be sure to fold up each batt or blanket at the ends of every joist space, as shown below, and abut any adjoining ends snugly.

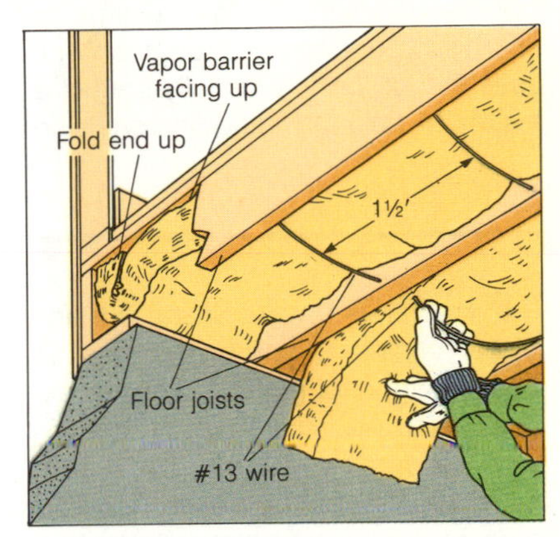

Stiff wire braces, cut 1 inch wider than the space between joists, hold insulation snug.

To form a vapor barrier in a dirt crawlspace, lay 4 to 6-mil opaque polyethylene sheeting over the ground, extending it several inches up the walls and fastening it there with duct tape. Overlap adjoining pieces and anchor them with rocks or bricks.

If the crawlspace can be sealed during winter, the cheapest and best protection is insulation draped across the inner surfaces of the exterior walls and along the ground, as shown on the facing page.

To insulate heated basement walls, you must either build a standard stud wall just shy of each concrete wall or fur out the concrete walls with smaller wooden framing members. Be aware, though, that because of the flammability of the vapor-barrier facings on regular blankets and batts, these materials cannot be left exposed. You must cover them with gypsum wallboard or some other fire-retardant covering permitted by local building codes.

And if you live in an extremely cold climate, you should check with your building department for recommendations before applying any type of insulation to basement walls.

Siding

Siding materials—wood, masonry, metal, or plastic—complete the exterior walls of your new structure. But new buildings aren't the only ones that need siding; it may be time to refurbish your present structure. What does the job entail? Usually it's just a matter of nailing siding materials to the wall studs or the present surface. Some materials require a backing of wall sheathing; others don't.

Two major siding materials—solid boards and plywood sheets—are discussed in detail in this section. Two popular alternatives, hardboard and plywood lap boards, are installed similarly. For tips on working with yet another choice, wood shingles or shakes, see pages 92–93.

Sidings such as aluminum, vinyl, steel, and stucco require special techniques or tools. If you're considering one of these products, get information from the manufacturer.

Preparing the Walls

The first step toward installing any type of siding is wall preparation. In new construction, wall studs are exposed and you're ready to begin the preparation steps outlined below. But if the wall is already covered with a layer of siding, you may have some additional work to do, such as adding furring strips or removing the old siding.

Sheathing. Structural sheathing is used under some siding materials to increase their rigidity, help brace the structure, serve as a solid base for nailing, and improve insulation. Check local building codes to determine whether sheathing is required with the siding you've chosen.

Sheathing is generally applied only to new structures. In most cases, existing siding on older walls acts as sheathing under a new covering.

Several types of sheathing are commonly used: plywood, exterior fiberboard, exterior gypsum board, and ordinary board lumber. Plywood is most popular because the large panels are easy to apply and usually afford enough lateral strength to eliminate the need for bracing during the framing of a house.

The chart below compares the main types of sheathing and basic application techniques. Nail all types directly to wall studs. As a rule, choose rustproof common nails; they should penetrate at least an inch into the studs.

WALL SHEATHINGS: HOW THEY RATE

	Types			
Qualities	**Exterior Plywood**	**Solid Boards**	**Exterior Fiberboard**	**Exterior Gypsum board**
Direction of application	Vertical or horizontal	Diagonal or horizontal	Horizontal	Vertical or horizontal
Panel sizes and types	5⁄16, 3⁄8, 1⁄2-inch thicknesses in panels of 4 by 8, 9, or 10 feet. Square-edge or tongue-and-groove.	1 by 6: end-matched tongue-and-groove. 1 by 8 or 1 by 12: shiplap.	1⁄2, 25⁄32-inch widths in 2 by 8-foot panels. Tongue-and-groove or shiplap.	1⁄2-inch widths, 2 by 8-foot panels. Tongue-and-groove.
Rigidity	Good	Good	Fair	Good
Insulative value	Low	Fair	Good	Low
Nailing	Nail every 6 inches along panel's edge and every 12 inches into center supports.	Three 8d nails per bearing for widths of 8 inches or more, 2 per bearing for lesser widths.	Use roofing nails 3 inches apart along edges, 6 inches apart intermediately.	Gypsum nails every 4 inches around edges and every 8 inches intermediately.
Does wall need diagonal bracing?	No	Only for horizontal application	Yes, with standard types	In some areas
General notes	Use performance-rated or exterior grade. Apply with panel ends spaced 1⁄16 inch apart and edges 1⁄8 inch apart.	Also available are 2 by 8 foot panels made from edge-glued lumber, overlaid with building paper.	Easy to handle and apply. Don't nail within 5⁄8 inch of edges. Only a special type will serve as sole nailing base for siding.	Not a nailing base for siding.

Building paper. Building paper is a wind and water-resistant material, usually felt or kraft paper impregnated with asphalt, applied between the sheathing or studs and the siding. It's purchased in rolls 36 to 40 inches wide, and long enough to cover 200 to 500 square feet (allowing for overlap).

Some local building codes require the use of building paper. You may also want to choose it as an option if your siding will be subjected to heavy winds or to wind-driven rain or snow, or if your siding consists of narrow boards or shingles that present many places for wind and water to penetrate.

Apply building paper in horizontal strips, starting at the bottom of each wall and working up. Overlap 2 inches at horizontal joints, 6 inches at vertical joints. Wrap the paper 12 inches around each corner. Cut building paper with a utility knife. Staple or nail it to studs or sheathing, using just enough fasteners to hold it in place. Later, the siding nails will fasten it permanently.

Removing existing siding. Avoid this job if possible. It's almost certain to be a lot of messy work, and it will expose your house to the weather. On the other hand, there may be no alternative if you want to install insulation batts or blankets inside existing walls. You also have no choice if your present siding is aluminum, vinyl, or steel, or if your existing siding—whatever the material—is in such bad shape that it *must* be removed.

Furring strips. If your present siding is masonry or is bumpy and irregular, you may need to provide a base of furring strips. Furring is generally a gridwork of 1 by 3 boards or strips, placed to provide nailing support for siding at the necessary intervals. Typical layouts are shown in the drawings at right. Nail through the strips every 12 inches with nails that penetrate studs at least 1 inch. If the existing walls are masonry, use concrete nails or masonry anchors to fasten the strips in place.

WEATHERTIGHT BACKING FOR NEW WALLS

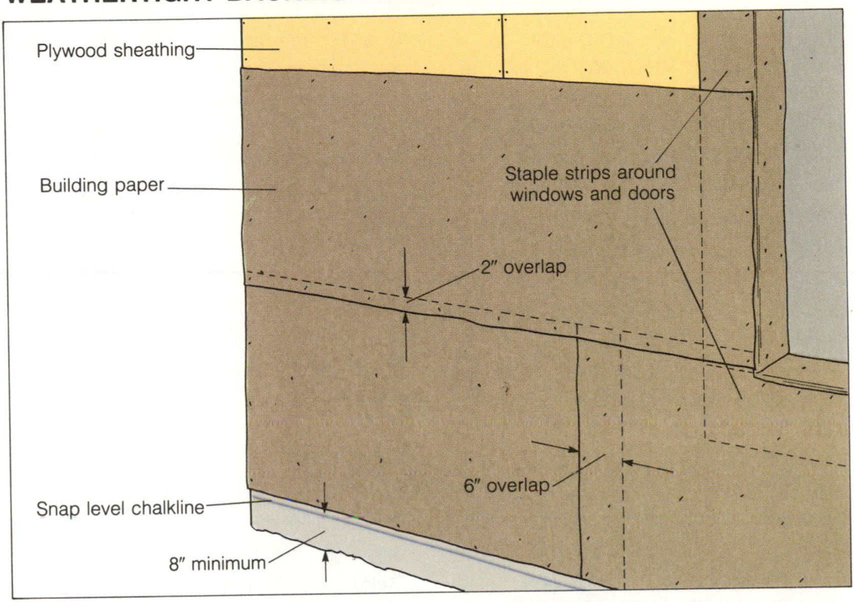

Plywood sheathing and building paper help seal this wall from the elements. Before siding goes on top, snap a level chalkline to indicate the bottom edge.

Base line and grading. No matter what type of siding you've chosen, you'll have to align its lowest edge along the base of each wall by snapping a level chalkline no less than 8 inches above grade (ground level). When new siding is going over old, the line is usually set 1 inch below the lower edge of the existing siding.

If necessary, excavate the surrounding soil where it interferes with this 8-inch clearance, sloping the grade away from the house so that water won't pool by the foundation. You may find it necessary to "step" the siding—adjusting the base line up or down—to conform to a hillside or irregular grade.

LAYOUTS FOR FURRING STRIPS

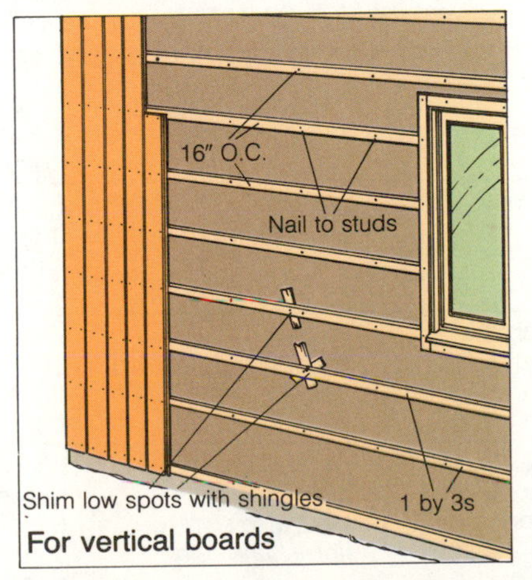

Furring strips smooth out a bumpy wall and provide a nailing base for siding materials. These two patterns can be adapted to most purposes.

. . . Siding

Installing Solid Board Siding

For the sake of simplicity, we've grouped all solid board siding patterns into a single category. But for proper installation, you must treat each basic pattern individually.

The chart below will tell you whether a particular pattern is applied vertically or horizontally, as well as the type of backing it requires, the size of nail to use, and the correct nailing technique.

Before you begin nailing up siding boards, figure out how you want to treat the corners. Typical treatments for both inside and outside corners are illustrated on the facing page. When planning your layout, try to plan board rows so that they fall evenly around windows, doors, and other openings. With horizontal siding, a slight adjustment to your base line may do the trick. If you must butt board ends together, stagger these joints as much as possible between successive rows.

Nailing. Solid board siding requires rustproof nails. Spiral or ring-shank nails offer better holding power than nails with smooth shanks; they work especially well in applying new siding over an existing wall covering.

The nail sizes in the chart are for new construction. When siding over an existing wall, choose nails that penetrate studs at least 1 inch. If you plan to countersink and fill over nailheads, use finishing nails.

Should your boards be so thin or dry that they tend to split, drill nail holes, especially at board ends. Blunting the tips of nails with a hammer is another way to help keep them from splitting boards.

Nailing the first board. With horizontal siding, the first board goes at the bottom. Beveled types require a starter strip beneath the board's lower edge, along the wall's base, to push it out to match the angle of the other boards (see drawing on facing page).

For vertical wood siding, begin at one corner of the house. Align one edge of the first board with the corner and check its other edge for plumb. If it isn't plumb, adjust it as necessary; then trim the outside edge with a plane or saw until it fits the corner. Be sure the board's lower end is flush with your base line; then nail it in place.

Successive boards. To lay out horizontal board siding, you'll need a "story pole." Make the story pole from a 1 by 3 that's as long as your tallest

SOLID BOARD SIDING: TYPES & INSTALLATION

Siding Type	Direction of Application	Nail Size (New Construction)	Nailing Tips
Board on Board (unmilled)	Vertical	8d underboards 10d overboards	For new construction, sheathing may be required. Otherwise, install blocks between studs on 24″ centers. Face-nail underboards once per bearing vertically; face-nail overboards twice, 3″ to 4″ apart, at center. Minimum overlap: 1″.
Board and Batten (unmilled)	Vertical	8d underboards 8d or 10d battens	For new construction, sheathing may be required. Otherwise, install blocks between studs on 24″ centers. Space underboards ½″ apart. Face-nail each board once every 24″ vertically. Minimum overlap: 1″.
Clapboard (unmilled)	Horizontal	10d	Face-nail 1″ from overlapping edge (just above preceding course) once per bearing. Minimum overlap: 1″. First board requires starter strip for correct angle.
Bevel, Dolly Varden	Horizontal	8d for ¾″, 6d for thinner	Face-nail once per bearing. With Dolly Varden, face-nail 1″ from lower edge. Allow expansion clearance of ⅛″. Minimum overlap: 1″. First board requires starter strip for correct angle.
Shiplap, Channel Rustic	Horizontal or vertical	8d for 1″, 6d for thinner	For vertical application, install blocks between studs on 24″ centers in new construction. Face-nail once per bearing for 6″ widths, twice (about 1″ from overlapping edges) for wider styles.
Tongue-and-Groove	Vertical, horizontal, or diagonal; can mix widths	8d (*finishing* for blind-nailing, otherwise *spiral* or *ring-shank nails*)	For vertical or diagonal application, install blocks between studs on 24″ centers in new construction. Blind-nail 4″ to 6″ widths through tongue with finishing nails, once per bearing. Face-nail wider boards with two spiral or ring-shank nails per bearing.

wall's height (unless that wall is more than one story). Starting at one end, mark the pole at intervals equaling the width of the siding boards.

Holding or tacking the story pole flush with the base line, transfer the marks to each corner and to the trim at each window and door casing.

Apply the siding boards from bottom to top. Unless the chart calls for overlapping or spacing, fit the boards tightly together. Where boards will be end-joined, brush a recommended sealant on the ends before installation and be sure to make the joint square and snug.

Where vertical boards join end to end, miter the joint at a 45° angle, with board ends sloping toward the faces, to ensure proper water runoff.

Two tips for problem spots. To fit a shiplap or tongue-and-groove board against a door or window frame, first rip it to the proper width. Then bevel the back edge that will butt against the frame; this will make it easier to push the board into place. To match the slope of a roofline, measure the angle with a T-bevel. Transfer the angle to each board end, then cut.

Soffits. Soffits (sometimes called cornices) are often left "open" when solid board siding is used—the boards extend to the tops of the rafters, as shown below, left. If you choose this method, you'll need to notch board ends where they intersect a rafter. Add trim along the top edge, between rafters, using molding or a narrow trim board (called a "frieze board").

If you prefer a "closed" soffit, you can buy or make a special wood board called a "plowed fascia board." This board, which is nailed over the rafter ends, has a routed groove near one edge for holding soffit boards or panels.

To complete the closed soffit, mark a point at both ends of the wall, level with the top of the groove in the fascia board. Snap a chalkline between your two marks; then nail a 2 by 2 above the chalkline. Measure the distance from the interior of the groove to the wall. Cut the soffit boards to fit, install them in the plowed fascia, and nail them to the bottom of the 2 by 2. Hide the nails with molding, fastening it with galvanized finishing nails set below the surface.

STARTER STRIP

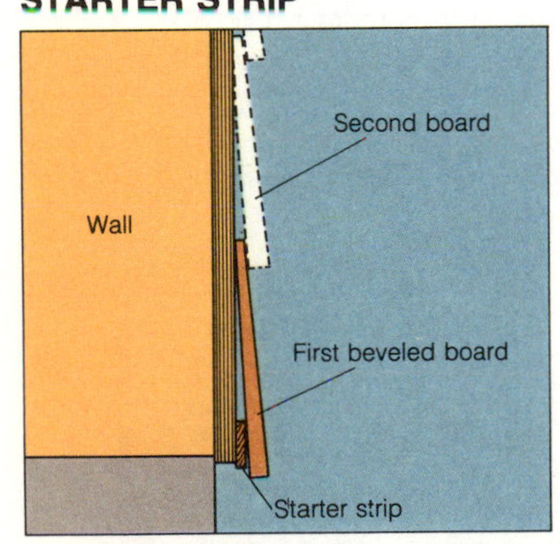

A horizontal strip props the first board out at the correct angle.

OPEN & CLOSED SOFFITS

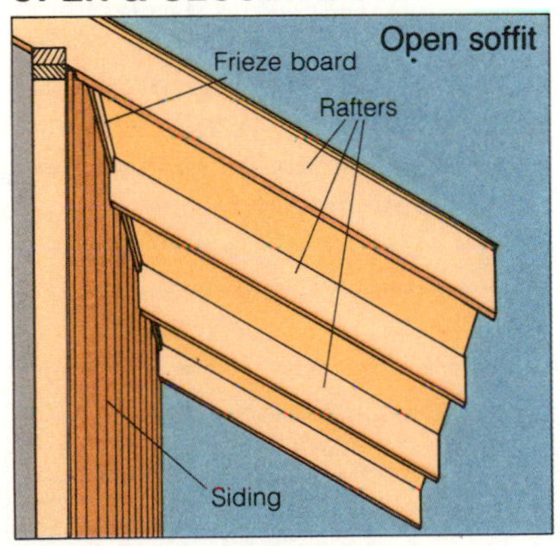

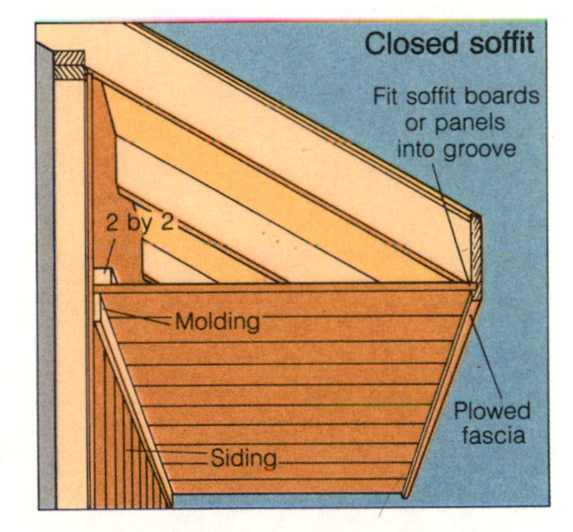

Soffits, the area below eaves, may be left open (left) or closed up with fascia and boards or panels (right).

CORNER TREATMENTS FOR SOLID BOARD SIDING

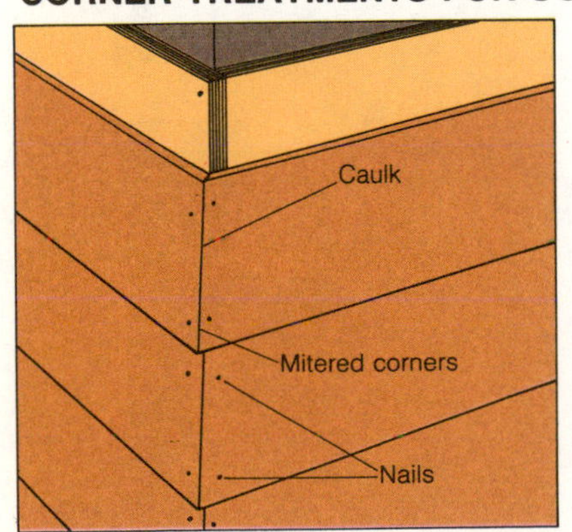

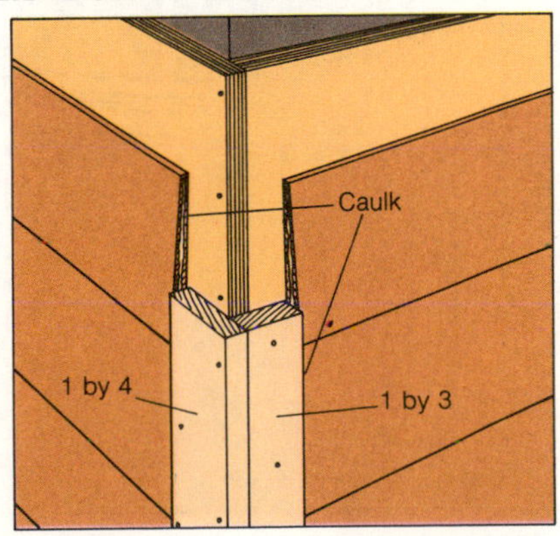

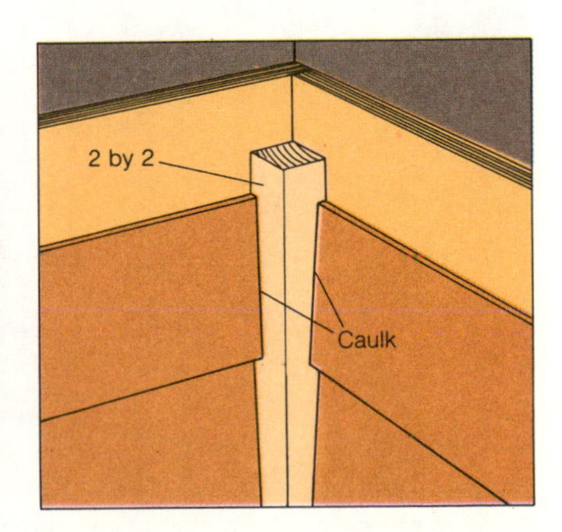

Where siding boards meet at corners, choose one of these three methods for moisture protection and a neat appearance.

. . . Siding

Installing Plywood Siding

Plywood's large panel size (4 by 8, 9, or 10 feet), its strong laminated construction, and its variety of surface styles make it a popular siding. Not only can it be applied rapidly, but it can eliminate the need for bracing and sheathing on the wall's frame.

Plywood siding applied directly to studs without sheathing must be at least 3/8 inch thick for studs on 16-inch centers, and at least 1/2 inch thick for studs on 24-inch centers. Panels as thin as 5/16 inch may be applied over wall sheathing or firm older walls.

Whatever pattern or style you choose, be sure to specify exterior-grade plywood.

Plywood panels may be mounted either vertically or horizontally. If you choose the horizontal pattern, stagger vertical end joints and nail the long, horizontal edges into fireblocks or other nailing supports. Because vertical installation minimizes the number of horizontal joints, it's the most common method.

Nailing. Nail plywood sheets with rustproof common or box nails. For re-siding over wood boards or sheathing, use hot-dipped galvanized ring-shank nails. Don't use finishing nails for plywood siding.

Nails should be long enough to penetrate studs or other backing by 1½ inches. For new siding nailed directly to studs, use 6-penny nails for 3/8-inch or 1/2-inch panels, and 8-penny nails for 5/8-inch panels. Nail every 6 inches around the perimeter of each plywood sheet, and every 12 inches along intermediate supports. When driving in nails, be careful not to "dimple" the plywood surface with the last hammer blow.

The first step. Before you begin putting up panels, you'll need to figure their correct lengths. They should reach from the base chalkline to the soffit. If you plan to create a closed soffit, it may be simpler to install it before the siding panels.

Should the distance from base line to soffit be longer than the plywood sheets, you'll need to join panels end to end, using one of the methods for horizontal joints shown on the facing page.

The first sheet. To begin your installation, position the first sheet at an outside corner, its bottom edge flush with the base line. Check with a carpenter's level to make sure vertical edges are plumb. If the corner itself isn't plumb, you'll need to trim the plywood edge to align with it. The inside vertical edge must be centered over a stud, furring strip, or other firm backing.

Hold or tack the sheet in place, flush with the base line, and trace along the outermost points of the existing siding or framing from top to bottom. Take the panel down and cut along this line with a circular saw or handsaw. Nail the trimmed panel in place.

NAILING UP PLYWOOD SHEETS

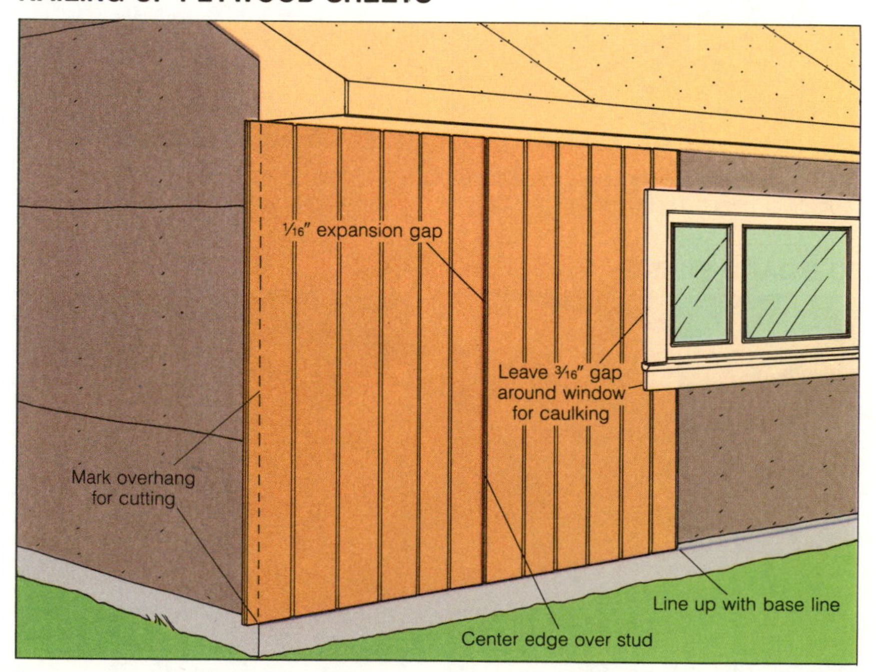

To hang plywood siding, start at one corner. Plumb, trim, and nail the first sheet, then butt the next sheet against the first.

OUTSIDE CORNERS

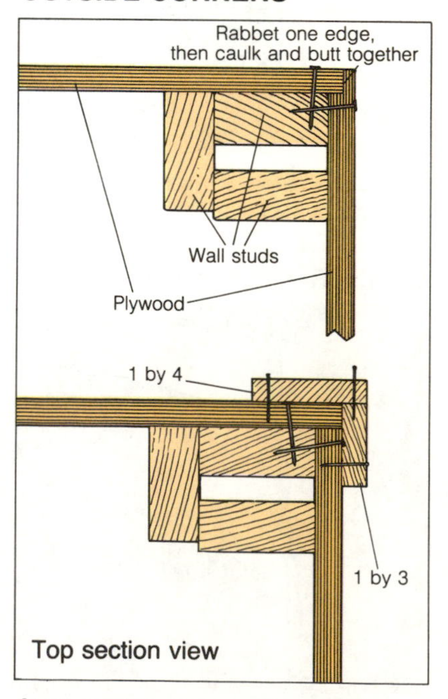

At an outside corner, either seal the joint with a rabbeted edge or add vertical trim boards.

Successive sheets. The next sheet butts against the first sheet, often with an overlapping shiplap vertical edge (see detail in drawing below center). Leave a 1⁄16-inch expansion gap at all joints (in humid climates, leave 1⁄8 inch). Sheets must join over studs, blocking, or other sturdy backing. Be careful not to nail through the laps.

If your plywood doesn't have a shiplap edge, caulk along the vertical edges and butt them loosely together, leaving about 1⁄16 inch for expansion. Unless there's building paper behind each joint, you'll need to cover the joints with battens (1 by 2 strips). All panel edges should be brushed with a recommended sealant before they're installed.

Corners. Plywood siding requires special corner construction to ensure a weathertight joint. Outside corners can either be rabbeted together and caulked, or covered by a 1 by 3 and 1 by 4 trim board, as shown on the facing page. Inside corners are generally just caulked and butted together. You can also supply a vertical corner trim board, such as a 2 by 2 (see drawing below left).

If the plywood is milled with grooves, the grooves under corner boards might let dirt and water penetrate. To prevent this, nail the vertical trim boards directly to the corner studs or siding, caulk along the edges, and butt the plywood against them.

Window and door openings. When cutting out these large areas, abide by the carpenter's maxim, "Measure twice, cut once." For ease of fitting, include an extra 3⁄16-inch gap around all openings. If possible, center the seams between sheets over or under the opening.

Use a carpenter's square or chalkline to lay out cuts, a circular saw for straight cutting, and a saber saw for cutting corners and curves. Since these saws cut on the upstroke, cut the sheets on the back side to avoid splintering the face. But beware: remember, when laying out the lines, that the sheet will be flipped over when installed.

Soffits. On plywood-sided houses, soffits are usually closed: an open soffit requires very exacting cuts to fit snugly around rafters. Consider running a plowed fascia board along rafter tails and lengths of siding along the soffit, as detailed on page 85.

INSIDE CORNERS

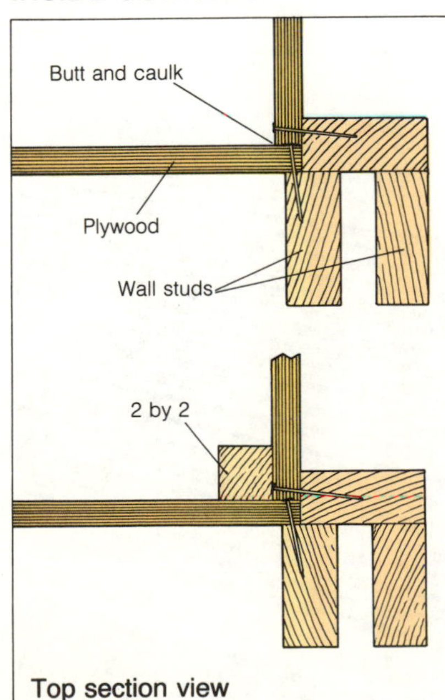

At an inside corner, simply butt and caulk sheets or add a vertical trim board atop the joint.

VERTICAL JOINTS

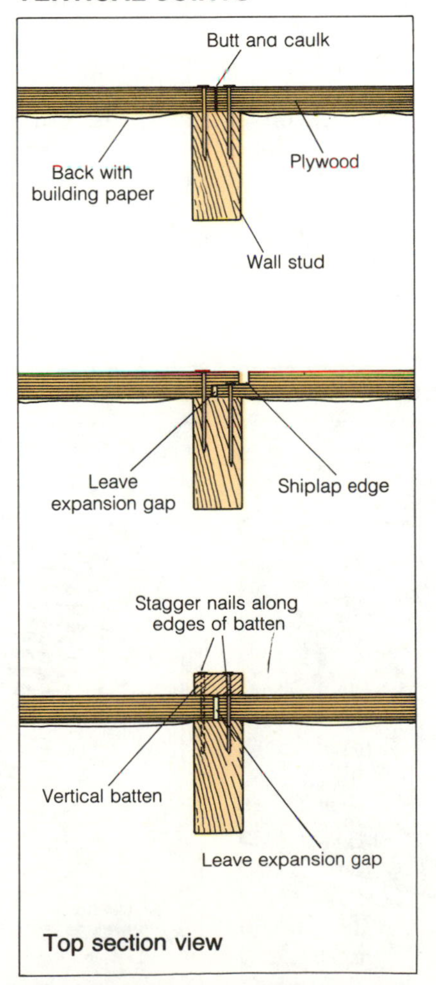

Where vertical panels meet, butt and caulk, or add a batten strip. Nail shiplap edges as shown.

HORIZONTAL JOINTS

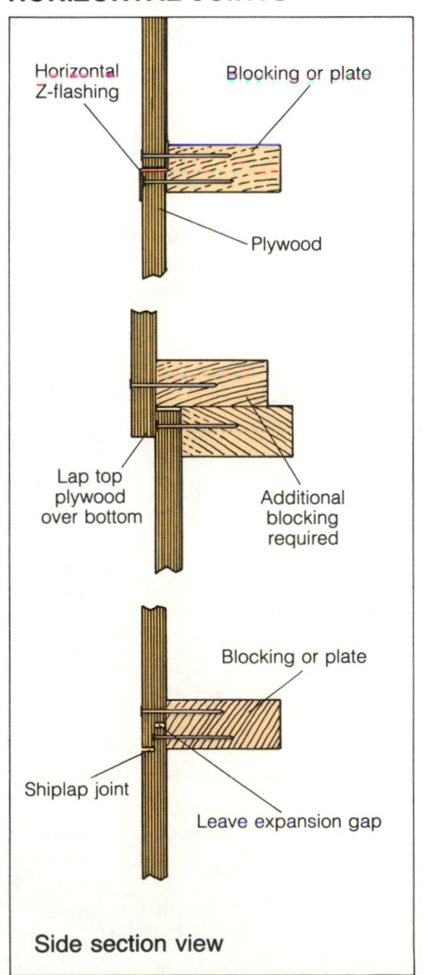

Protect horizontal seams with metal Z-flashing, by overlapping, or by choosing panels with shiplap edges.

Roofing

Preparations for a new roof begin with a smooth deck atop the rafters followed by underlayment and then by protective flashings at spots especially vulnerable to moisture. After that, you're ready to begin actual roofing.

We've included instructions for roofing with asphalt shingles, wood shingles, and wood shakes—today's most popular choices. Whatever surface you choose, the roofing procedure boils down to a three-stage operation: first you lay a "starter" course to bolster the initial row of materials; second, you apply materials row by row, from eaves to ridge; and third, you cap the roof's ridges and hips with specially made shingles.

Can you lay a new roof right on top of your old one? If the present roof is asphalt, and both roof and sheathing are in good condition, probably so. You can also lay a new roof over wood shingles, though many professional roofers advise against reroofing over either wood shingle or shake roofs. Whatever the material, if you're dealing with a badly worn surface or rotted sheathing, or if your home has already had the maximum number of reroofings permitted by code, you'll have to tear the old roof off.

Decking for New Roofs

The type of material you'll choose for roof sheathing or *decking* will depend on the type of roof you plan to install. Asphalt shingles should be applied over a solid deck of plywood sheathing or boards, with an underlayment of 15-pound roofing felt. Wood shakes, too, are often laid over solid decking, though in many instances they're placed over spaced 1 by 4 boards. Wood shingles are typically laid atop spaced decking.

Installing "solid" plywood decking. When installing plywood sheathing, stagger panels horizontally across rafters, with ends centered on the rafters. To allow for expansion, leave a 1/16-inch space between the ends of adjoining panels, and 1/8 inch between edges. If your climate is exceptionally humid, double this spacing.

Though some codes permit the use of plywood as thin as 5/16 inch on roofs with 12-inch rafter spans, or 3/8 inch thick on 24-inch spans, you'll probably want either 1/2 or 5/8-inch sheathing for a sturdier nailing base. If you opt for the lighter panels, you may need tongue-and-groove edges, plywood sheathing clips, or solid blocking for support between rafters.

When nailing plywood to rafters, use 6-penny common or box nails for plywood up to 1/2 inch thick, or 8-penny nails for plywood 5/8 inch and thicker. Space nails every 6 inches along the vertical ends of each panel, and every 12 inches at intermediate supports.

Rolling out the underlayment. To evenly align rows of underlayment, measure the roof carefully and snap horizontal chalklines before you begin. Snap the first line 33 5/8 inches above the eave (this allows for a 3/8-inch overhang). Then, providing for

OPEN & CLOSED: TWO ROOF DECKS

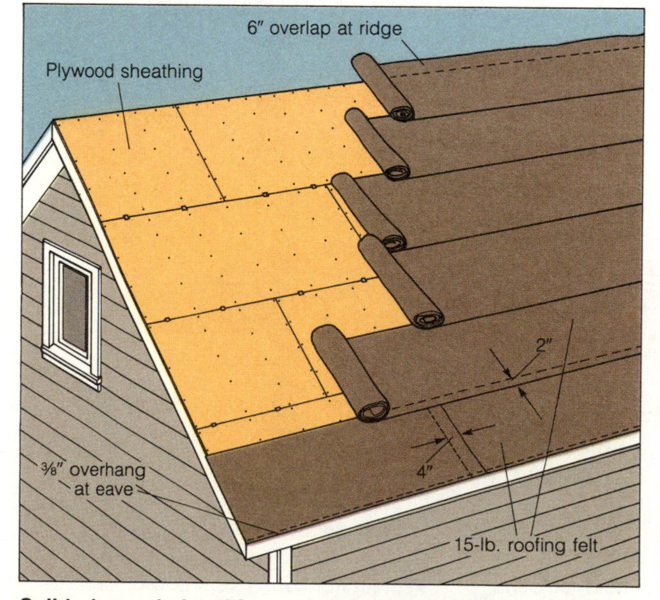

Solid plywood sheathing and roofing felt form a closed deck for asphalt shingle or wood shake roofs.

Spaced 1 by 4 boards form an open deck for wood shingles. At eave and rake overhangs, close the gaps with extra boards.

a 2-inch overlap between strips of felt, snap each succeeding chalkline at 34 inches.

When applying felt, start at the eave and lay the strips horizontally along the roof, working toward the ridge. Felt should be trimmed flush at the rake (gable overhang) and overlapped 6 inches at any valleys, hips, and ridges. Where two strips meet in a vertical line, overlap them by 4 inches.

Drive just enough nails to hold the felt in place (generally, a 1¼-inch galvanized roofing nail for every square yard of felt).

Installing "open" decking. For spaced sheathing, lay well-seasoned 1 by 4 boards horizontally along the roof, using another 1 by 4 as a spacing guide. Fasten each board to the rafters with two 8-penny nails, allowing ⅛-inch spacing where boards meet (see drawing on facing page).

At overhangs, you'll want continuous sheathing—without gaps. Start your installation with solid rows of 1 by 4s at the eaves and rakes, as shown.

Roof Flashings

Flashings protect your roof at its most vulnerable points: in the valleys (see drawing below); at the edges of roof vents, chimneys, and skylights; along the eaves of the house—anywhere water can seep through broken joints into the sheathing. Flashings are most commonly made of a malleable, 28-gauge galvanized sheet metal. On asphalt shingle roofs, valleys and vents may also be flashed with mineral-surface roll roofing. Plastics or aluminum are used, too, and copper is sometimes preferred for chimney flashing. You can either buy preformed flashings for drip edges, valleys, or vents, or make your own.

Valleys require particularly sturdy flashing because they conduct more water to the gutters than do any of the roof planes. Finishes vary from "open" valleys, where shingles are cut away to expose the valley, to "woven" valleys, where asphalt shingles overlap the valley.

Drip edges are required at eaves and rakes. Along eaves, install drip edges underneath the roofing felt. Along rakes, install them on top.

Vent pipe flashings are usually installed with the roofing materials, since part of the flashing fits over the shingles.

Chimney flashing typically consists of a solid base flashing along the bottom, overlapping step flashing up the sides, and a continuous saddle flashing at the top. A second layer, or *cap* flashing, is often added to the first.

Vertical wall flashing protects the area where a sloped roof meets a vertical wall. You can install L-shaped metal that's crimped at the edges if it will later be covered by siding. If not, you'll need to install a 12-inch-wide strip of metal or asphalt roll roofing along the intersection, then caulk the seam where the strip meets the vertical wall.

FLASHINGS FOR FIVE TROUBLE SPOTS

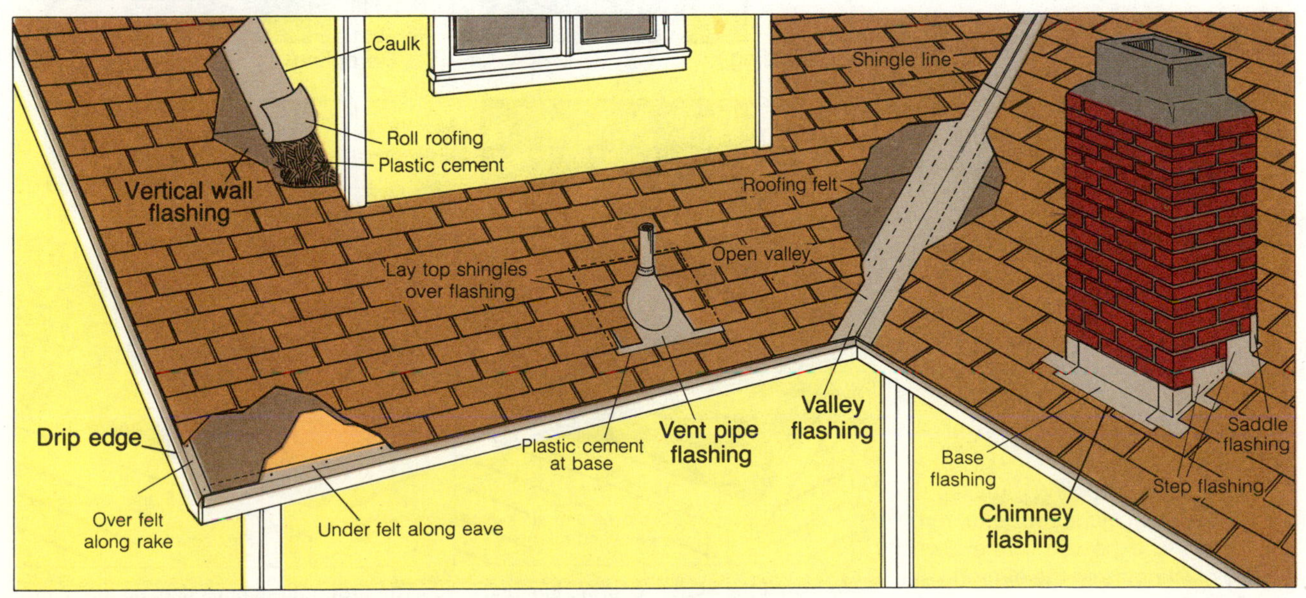

This roof flashing sampler shows the main types and their major components. Check building codes for exact requirements in your area.

...Roofing

Applying Asphalt Shingles

Asphalt shingles come in a wide variety of colors, from subtle earth tones to crisp whites, reds, greens, and blacks. Standard three-tab shingles (see drawing below) measure 12 inches by 36 inches. Most have a self-sealing mastic that welds one shingle tab to another after the shingles are installed.

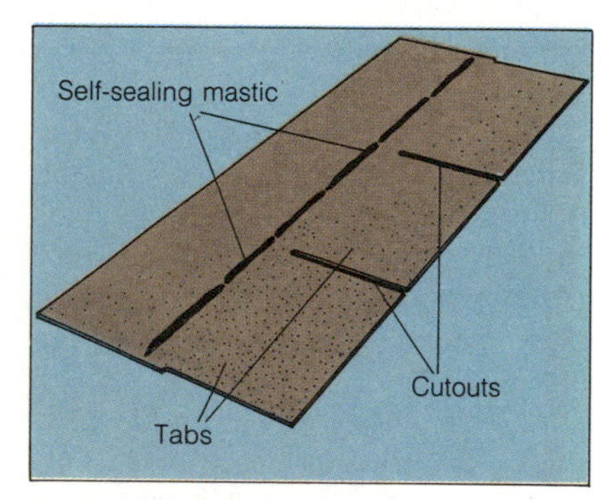

The standard asphalt shingle is 12 inches by 36 inches and has three integral tabs.

The correct weather exposure for most asphalt shingles is 5 inches. That means that the lower 5 inches of each shingle will be exposed to the weather after overlapping courses are applied.

Nails and nailing. For asphalt shingles, choose 12-gauge, galvanized roofing nails with ⅜-inch-diameter heads. New asphalt roofs need 1¼-inch-long nails. Choose 1½-inch nails when roofing over an old asphalt roof, and 1¾-inch nails when roofing over an old wood roof.

Starter course and first course. The narrow starter row of shingles runs the length of the eave to form a base for the first full course of shingles. Before laying the starter course, first measure the eave and then select enough 36-inch-long shingles to cover the distance.

If you're reroofing, make the starter course 5 inches wide to correspond to the exposure of the lowest course of shingles on the old roof (see drawing at right). Cut 5 inches off the tabs and 2 inches from the top edges of 12-inch-wide shingles. To cut shingles, use a utility knife and carpenter's square or straightedge. Score the backs of shingles and then bend them to break.

For a new roof, apply a 9-inch-wide starter course. Either use a 9-inch strip of asphalt roll roofing or cut 3 inches off the tabs of 12-inch-wide shingles.

Starting at the left rake, apply the starter course along the eave with the shingles' self-sealing strips down. Trim 6 inches off the first shingle's length to offset the cutouts in the starter and first courses.

Allowing a ½-inch overhang at both eave and rake, and 1⁄16-inch spacing between shingles, fasten the shingles to the deck using four nails placed 3 inches above the eave. Position these nails 1 and 12 inches in from each end.

To lay the first course of a new roof, use full-width shingles. When reroofing, you'll need a 10-inch-wide course to cover the two 5-inch exposures of the first two courses of old shingles. Cut 2 inches off the top edges of as many shingles as were necessary for the starter course.

Allowing the same ½-inch overhang at the rake and eave, and 1⁄16 inch between shingles, nail the first course over the starter course, using four nails per shingle. Space these nails 5⅝ inches above the butt line, 1 and 12 inches in from each end (or according to manufacturer's instructions).

Successive courses. Your main concern when you lay the second and successive courses is proper alignment of the shingles—both horizontal and vertical. Aligning shingles horizontally is simply a matter of snapping chalklines across the deck (new roof) or placing new shingles against the butts of old ones (reroofing). If you're using chalk, snap lines every 10 inches from the bottom of the first

STARTER COURSE, FIRST COURSE

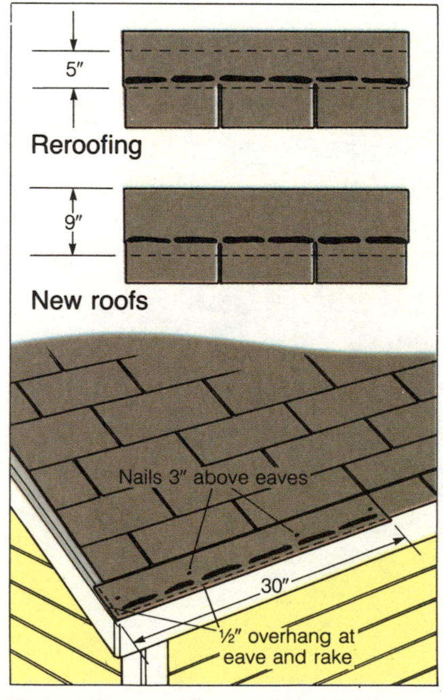

To install the starter course, trim and nail shingles from rake to rake along the eave.

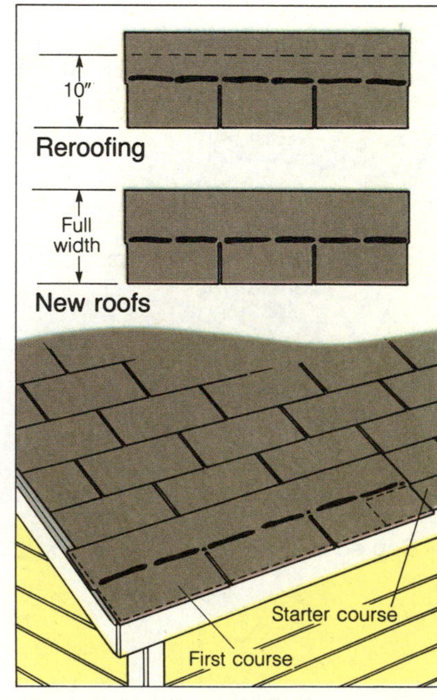

The first course goes right on top of the starter course, with tabs offset by 6 inches.

course, as the drawing below indicates. Then, as you move toward the ridge, the upper edge of every other course of shingles should line up against the chalklines.

Before you start your second row of shingles, also snap vertical chalklines from the roof ridge to one end of every shingle along the first course.

If you're working with standard three-tab shingles, you can produce centered, diagonal, or random roof patterns simply by adjusting the length of the shingle that begins each course.

Centered alignment offers the most uniform roof appearance, but it's also the most difficult pattern to achieve because cutouts or shingle edges must line up—within ¼ inch—with those two courses above and below (see drawing below).

Diagonal alignment is a little more forgiving if you make slight errors in your calculations.

Random alignment is the easiest of the three patterns to lay; it produces a more rustic appearance in the roof surface.

Installing hip and ridge shingles. If you haven't purchased ready-made ridge and hip shingles, cut 12-inch squares from standard shingles. Bend each square to conform to the roof ridge. (In cold weather, warm shingles before bending.) Then, before you begin, snap a chalkline the length of the ridge and each hip, 6 inches from the center.

If your roof has hips, start with them. Beginning with a double layer of shingles at the bottom of the hip, work toward the ridge, applying shingles with a 5-inch exposure. The edge of each shingle should line up with your chalk mark.

Use two nails, one on each side, 5½ inches from the butt and 1 inch from the outside edge (see drawing).

To shingle the ridge, start at the end opposite the prevailing wind. Using nails long enough to penetrate the ridgeboard securely (about 2 inches long), apply the shingles in the manner just described. Dab the exposed nailheads of the last shingle with plastic cement.

THREE PATTERNS FOR ASPHALT SHINGLES

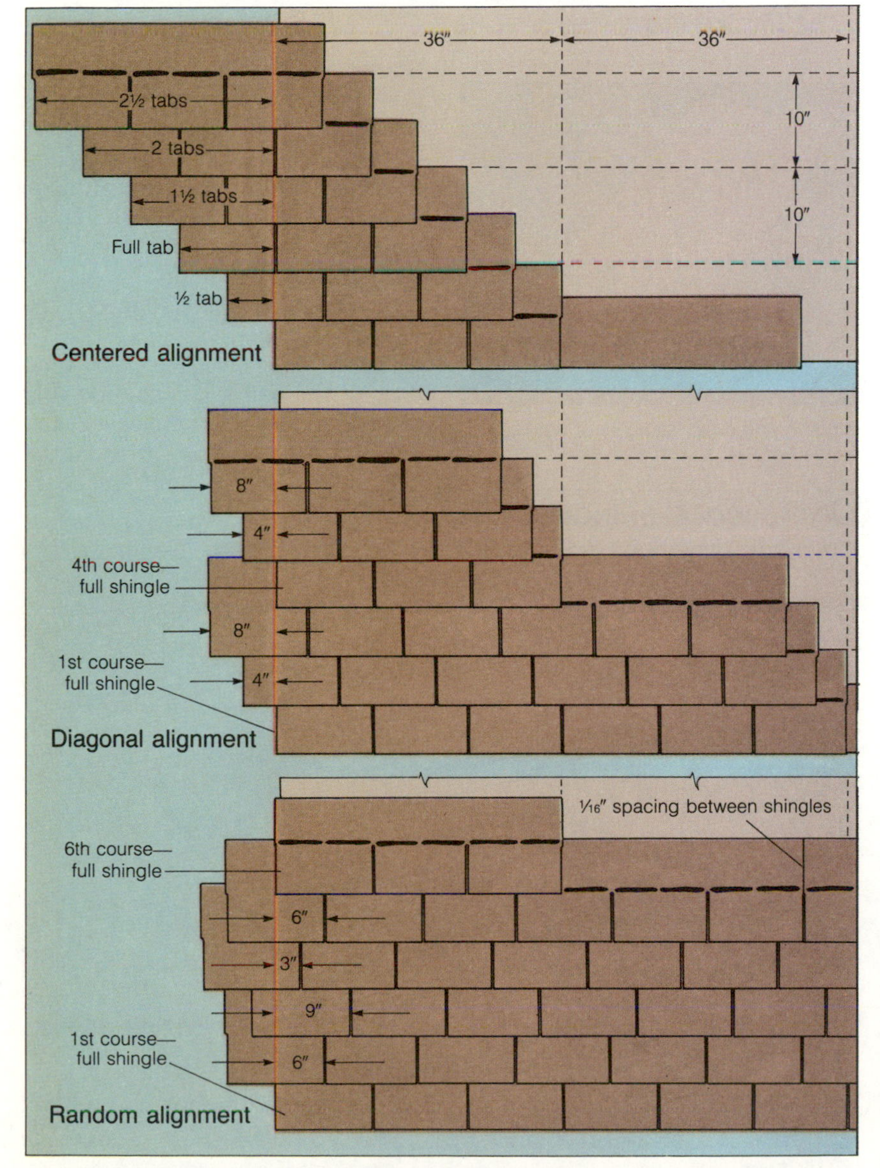

To lay out new shingles, snap chalklines or follow existing courses, choosing one of these three patterns. Simply adjust the length of the shingle beginning each course to determine the pattern.

Cover hips and ridges with specially made shingles, or cut and bend your own.

. . . Roofing

Applying Wood Shingles & Shakes

Shingles, with their smooth, finished appearance, are sawn from chunks (called "bolts") of western red cedar. They come in lengths of 16, 18, and 24 inches. Shakes are thicker, and instead of being sawn from the bolt, they are split by machine or by hand into 18 and 24-inch lengths. Though shingles and shakes are available in several grades (suitable for siding as well as roofing), you should be sure to specify Number 1 ("Blue Label") grade for use on a roof.

When applying either shingles or shakes, always position the tapered end uproof and the thicker end downroof. If the wood has a sawn side and a rough side, install with the rough side exposed to the weather. When applying straight-split shakes (those equally thick throughout), lay them with the smooth end uproof.

Exposure. Correct exposure for wood shingles and shakes depends on their length and the slope of your roof. Here are recommended exposures.

MAXIMUM EXPOSURE SHINGLES & SHAKES

	3 in 12 to 4 in 12 slopes	4 in 12 and steeper slopes
Shingles:		
16"	3¾"	5"
18"	4¼"	5½"
24"	5¾"	7½"
Shakes:		
18"	—	7½"
24"	—	10"

Shingle exposures are for Number 1 (Blue Label) shingles only.

Nailing. For wood shingles and shakes, use rustproof nails, two per shingle or shake.

Nails for wood shingles should be 14½-gauge with 7/32-inch-wide heads. Use 1¼-inch-long nails for a new roof of 16 or 18-inch shingles; use 1½-inch nails for a new roof of 24-inch shingles. Longer nails may be needed to penetrate old roof surfaces and reach ¾ inch into or through the deck. Above roof overhangs, choose shorter nails.

For wood shakes, the preferred choice is 13-gauge nails with 7/32-inch-wide heads. Use 2-inch long nails unless longer ones are required to penetrate ¾ inch.

When fastening shingles, locate the nails ¾ inch in from each side and 1 inch above the butt line for the next course. Nails for shakes are positioned 1 inch in from the sides.

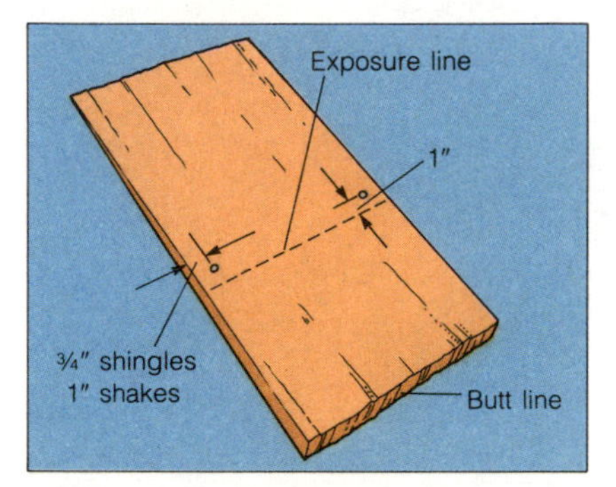

When laying down shingles or shakes, space nails as shown above.

How to trim shingles or shakes. To make straight cuts in shingles, simply slice through them with a roofer's hatchet. Heavier wood shakes can either be sawn or split along the grain with the hatchet.

When it's necessary to make an angled cut for a valley, lay the shingle in place and use a straightedge to mark the angle of the cut. Then score the shingle with a utility knife and break it against a hard edge. Wood shakes must be sawn.

Starter course and first course. Combine the starter and first courses by laying the shingles or shakes one on top of the other (see drawings below and on facing page). Though roofing felt is seldom required under wood shingles, it is recommended with shakes; their irregular shape allows water to work through the cracks.

So if you're roofing with wood shakes, first nail a 36-inch-wide strip of 30-pound felt along the eave (allow a ⅜-inch overhang).

LAYING WOOD SHINGLES . . .

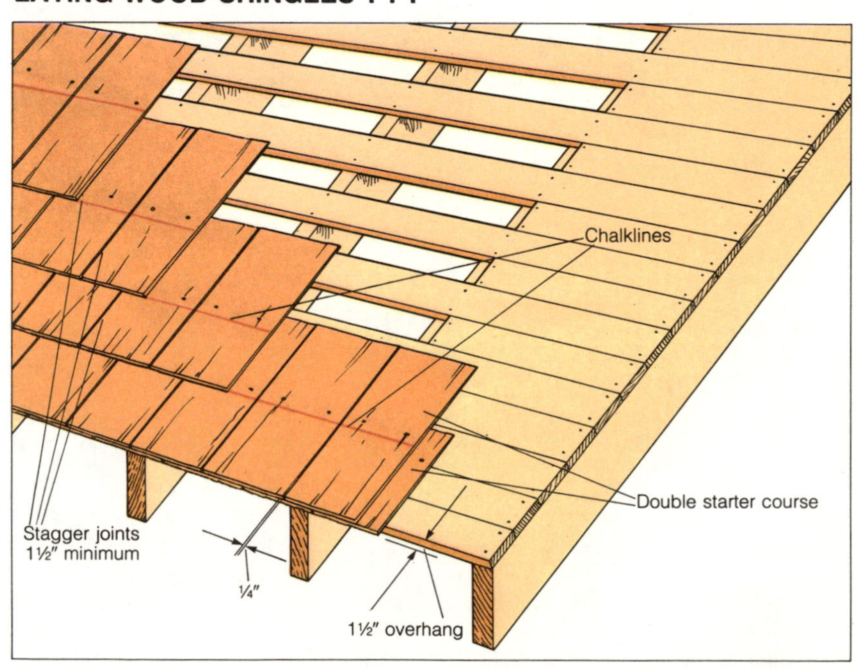

Apply wood shingles in even courses, snapping chalklines to indicate the correct exposure. Nails must be covered by the butts of each new course.

When laying the double course, offset the joints between layers by 1½ inches. This double course should have a 1½-inch overhang at the eaves and rakes. Allow spacing of ¼ inch between shingles and ½ inch between shakes to permit the wood to expand and contract.

Successive courses. When you lay the next courses, align shingles or shakes both vertically and horizontally for proper exposure and coverage.

You don't need to snap chalklines when you line up shingles or shakes vertically. Simply lay the random-width wood materials according to this principle: offset joints by at least 1½ inches so that no joints in any three successive courses are in alignment.

To line up wood shingles horizontally, snap a chalkline at the proper exposure over the doubled starter/first course, or use your roofer's hatchet as an exposure guide. Then lay the butts of the next shingle course at the chalkline. Nail the course down and repeat the procedure until you reach the ridge.

When aligning wood shakes, install roofing felt interlays over each course as you work toward the ridge. Here's how: From the butt of the starter/first course, measure a distance twice the planned exposure. Place the bottom edge of an 18-inch-wide strip of 30-pound felt at that line, and nail every 12 inches along the top edge of the felt. Overlap vertical joints 4 inches.

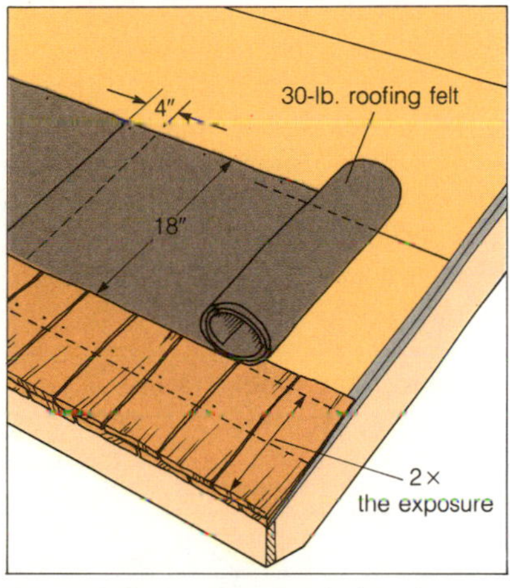

To install felt interlays for shakes, lay an 18-inch strip between each successive course.

. . . AND WOOD SHAKES

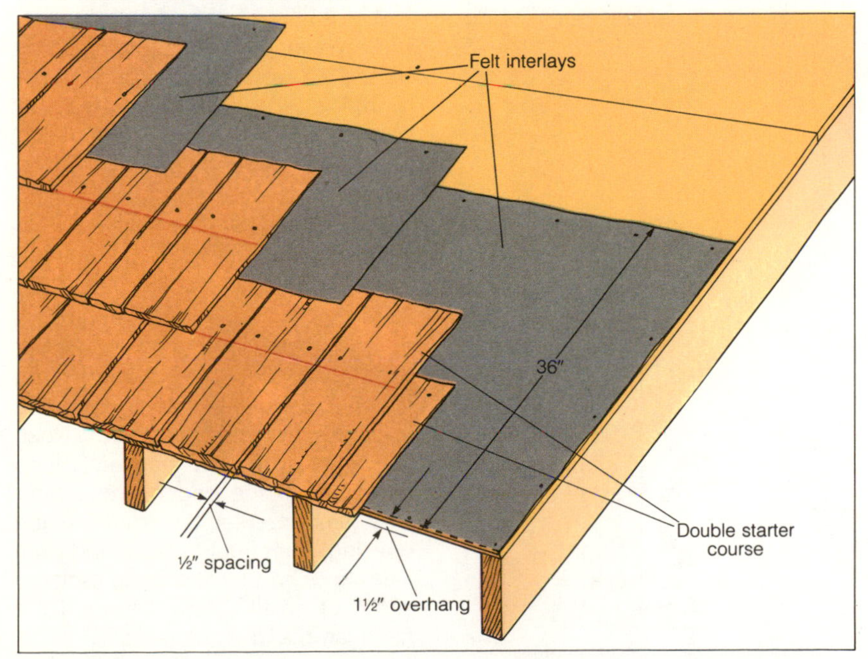

Wood shakes are put down much like shingles, but roofing felt is required for extra moisture protection. Be sure to leave ½-inch spacing between shakes.

Then snap a chalkline on the starter/first course for the proper exposure (or use your roofer's hatchet for a guide). Nail the second course, place the next felt, and continue until you reach the ridge.

Use short, 15-inch shakes as the last course. They can either be ready-made or cut from standard shakes.

At the ridge, let the last courses of shingles or shakes hang over; snap a chalkline above the center of the ridgeboard and trim off all the ends at once. Cover the ridge with a strip of 30-pound felt at least 8 inches wide.

Applying hip and ridge shingles. Using factory-made ridge and hip shingles (mitering and making your own specialty shingles is a time-consuming task), double the starter courses at the bottom of each hip and at the end of the ridge, as indicated in the drawing below.

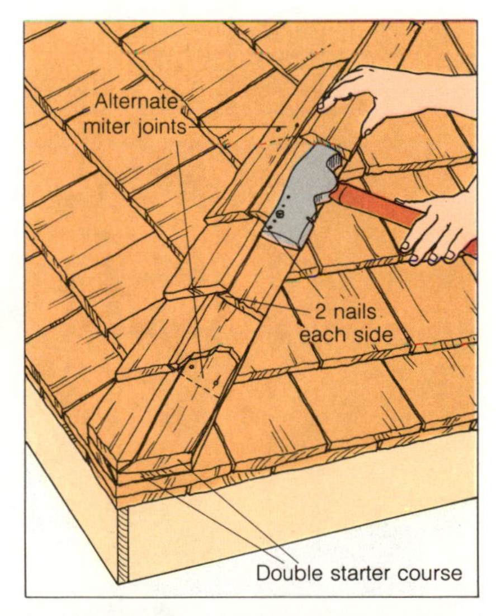

Lay hip shingles with the same exposure as shingles on the roof planes; work up from the double starter course. A roofer's hatchet has a handy exposure gauge.

Exposure should equal the weather exposure of the wood shingles or shakes on the roof planes. Start the ridge shingles at the end of the ridge opposite prevailing winds. Use nails long enough to penetrate the layers of material and extend into the ridgeboard (usually 2 or 2½ inches).

Doors

Installing a door the traditional way requires a bit of patience as you construct a door frame, position it on the wall inside the rough door opening, hinge the door to the frame, and install a latch or lockset. By purchasing a prehung door unit, you can bypass several steps—though you'll end up with a less customized job. The choice is yours; both procedures are described below.

Before you're ready to install a door, of course, you must frame the rough opening. If you're building a new structure, you'll frame the openings along with the walls—see "Wall Framing," pages 70–71. Remodeling calls for a different approach: after removing wall materials and studs, you frame the opening with the wall in place. For details, turn to page 96.

Hanging a Door—From A to Z

Modern manufactured doors come in two basic types: *panel* and *flush*. Panel doors, like the one shown below, consist of solid vertical stiles and horizontal rails, with filler panels between. Superior strength and good looks make this style a wise choice for exterior doors, especially at a main entrance. Flush doors, on the other hand, are built from thin face and back veneers—typically ⅛-inch plywood—attached to a solid or gridlike hollow core. The hollow-core type is for interior use only.

Standard door height is 6 feet 8 inches. Width and thickness vary: for exterior doors, choose a width from 2 feet 8 inches to 3 feet, and a thickness of 1¾ inches. Interior doors are typically 2 feet 6 inches wide and 1⅜ inches thick. Both types are installed in a similar manner, with a few extra steps required for weatherproofing an exterior door.

Installing the door frame. A door frame consists of two side jambs and a head jamb. The tops of the side jambs are normally grooved to hold the head jamb. You can buy standard jambs at lumberyards. Choose a jamb width equal to the thickness of the finished wall. For example, if the wall is framed with standard 2 by 4 lumber and surfaced on both sides with ½-inch wallboard, your jambs should be 4½ inches wide. Jambs for exterior doors may include integral stop moldings (see below).

The width of your frame is determined by the length of the head jamb. To decide what this length should be, add up the following: the width of the door to be installed; 1⁄16-inch clearance between the door and each side jamb (⅛ inch total); and the depth of the side jambs' routed grooves. Cut the head jamb to this length, and then assemble the frame with glue and finishing nails.

An exterior door needs a sloping sill and a threshold at the base of the jamb. The sill is installed flush with the finish floor, so you may need to notch out the subfloor and possibly the joist header or stringer below. Sill or no sill, you'll need to cut out the length of the sole plate between trimmer studs (see page 71).

Prop up the jamb assembly in the opening; if the finish floor is not yet installed, insert scraps of the flooring material or small blocks of the same thickness below the side jambs. Next, check the level of the head jamb. If it's uneven, trim the end of the higher side jamb. Center the frame in the opening, from front to back and from side to side, by wedging shingle shims or small blocks between the head jamb and header.

Now, beginning next to the lower hinge location, drive two shims snugly

ANATOMY OF A DOOR

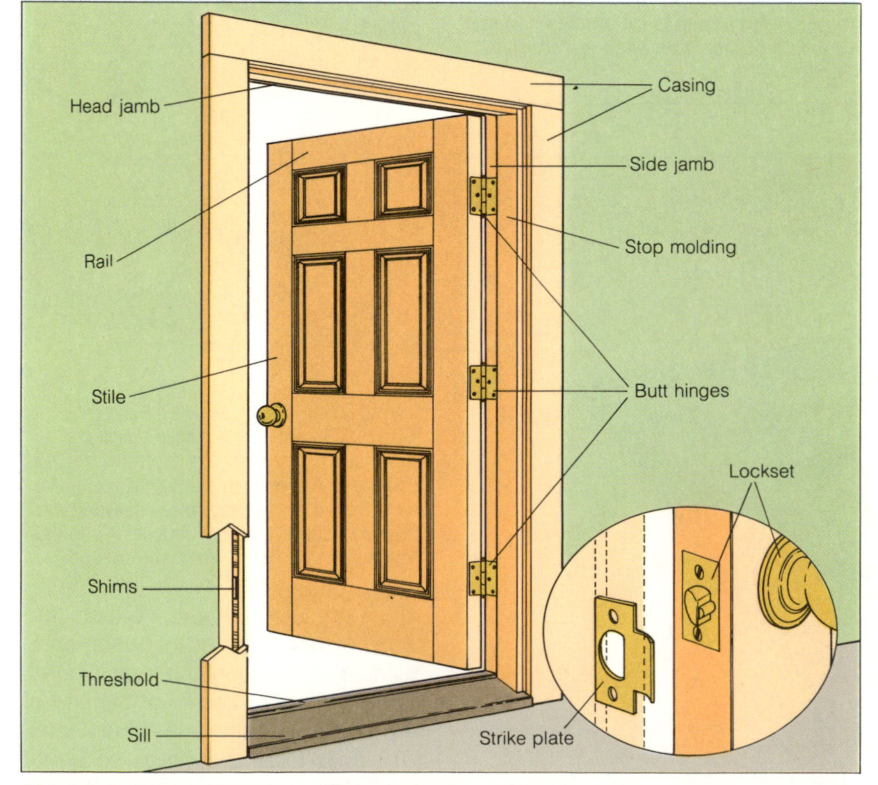

This panel door is hinged to a frame formed by a head jamb and two side jambs. Door casing seals the frame and ties the structure together. An exterior door needs a sill and threshold at the bottom.

between the side jamb and trimmer stud—one shim driven from each side. Nail through the jamb and shims partway into the stud with an 8-penny finishing nail; be sure to position the nail where the stop molding (see drawing on facing page) will cover it. Insert shims next to the upper hinge location, check the jamb for plumb, and nail partway. Again, shim, plumb, and nail halfway between top and bottom hinge positions.

Now shim the opposite jamb at similar locations; locate the middle shims behind the strike plate, but don't nail where you'll need to notch or bore.

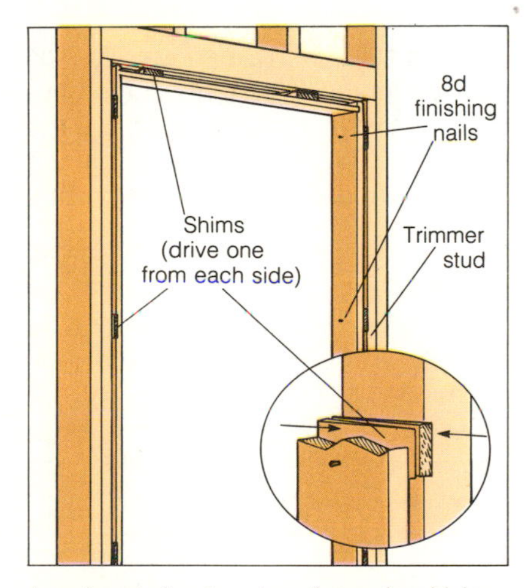

Level and plumb a door frame by driving wood shims between jambs and rough framing.

Check the frame once more for plumb. When all is in order, drive the nails home and set their heads with a nailset. If you're working on an exterior door, nail the threshold between the jambs (unless threshold and sill are a single unit).

Attaching the casing. Door casings may be either contoured moldings or standard lumber. If you choose lumber, plan to butt joints together at the top; for molding, you'll have to miter the tops. See page 111 for tips on installing casing.

Installing the hinges. For most interior doors, hinges should be placed about 7 inches from the top and 11 inches from the bottom of the door. Exterior doors and heavy interior doors require a third hinge, centered between the first two. Choose 3½-inch-long butt hinges for interior doors and 4½-inch hinges for exterior ones.

To lay out hinge locations, prop the door up within the frame, raising it on blocks to the correct height above the floor. (A 4-penny nail will give you the correct top spacing.) Mark the hinge locations on door and jamb simultaneously.

Remove the door and cut mortises for the hinges with a sharp chisel (as shown on page 14) or an electric router and template (pages 16–17). Be sure to leave ¼ inch between the hinge edge and the back of the door. The hinge must sit flush with the top of the door.

Attach the matching hinge leaves to the door jamb in the same manner and you're ready to hang the door. Place the pin in the top hinge first, then in the bottom.

Door troubleshooting. Does the door bind slightly at the top or bottom of the lock side? You can offset this problem by inserting cardboard shims behind the hinge leaves. If the door sticks at the top, shim the lower hinge; if at the bottom, shim the top hinge. To correct a clearance that's too tight at top *and* bottom, insert a shim opposite the pin side of each hinge, as shown below. Too loose? Insert the cardboard on the pin side.

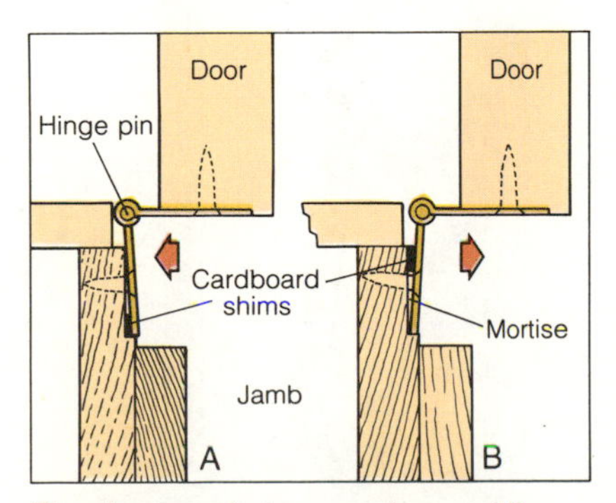

The clearance between a door and the strike plate jamb can be increased (A) or decreased (B) by shimming between the opposite jamb and hinge leaf.

If your door is much too tight on any side, you'll have to scribe the correct clearance along the door, remove it, and shave the door to the line. For small adjustments, use a jack plane or belt sander; cuts over ¼ inch require a circular saw.

Installing a lockset. A template to help you position holes for the lock, latch, and strike plate should be included with your lockset. Place the knob 36 to 38 inches above the floor. If you're using a hand brace, you'll need an expansion bit and auger bit to bore the holes; with an electric drill, choose a hole saw and spade bit. The lockset should come with detailed instructions for installation.

To install a lockset, you'll need to mark and drill holes for both lock and latch.

After installing both the lock and the latch, close the door and mark the top and bottom of the latch where it contacts the jamb. Position the strike plate there; drill or chisel out a mortise for the latch.

Finishing touch: Stop molding. Unless your door frame includes them, you'll need to add a narrow strip of stop molding along each jamb to prevent the door from closing too far. Beginning at the hinge side, nail a length of molding from the floor to the head jamb with one 4-penny finishing nail every 12 inches, spacing it ¹⁄₁₆ inch from the door face so the door won't bind. Nail the opposite stop flush with the door and connect the two across the head jamb. If the door binds, bevel the inside edge on the hinge side to help it clear the stop molding.

...Doors

Installing a Prehung Door

Once you become acquainted with the prehung door unit, you'll discover that much of the tricky work associated with door hanging has been done for you: the door comes hinged to the jambs, and frequently casing, lock, latch, and strike plate are included, too.

To complete the installation, first remove any bracing, blocking, or stop molding tacked to the unit. Position the unit in the rough opening. If the finish floor is not yet installed, raise the side jambs to the correct level with blocks.

Shim the jambs level and plumb, as detailed on page 95. If casing is included on the face side, insert shims from the back only. Fasten the unit to the rough door framing by driving 8-penny finishing nails through the jambs and shims. Fasten the attached casing to the rough door framing near the outside edge with 8-penny finishing nails spaced every 16 inches.

Check the door for adequate clearance and smooth operation. To make adjustments, see "Door troubleshooting" on page 95. If the unit did not include a lockset, install it now according to the manufacturer's instructions.

Add casing to the back side of the door. Finally, nail the stop molding to the jambs—1⁄16 inch back from the door along the hinge side, and flush along the head and strike plate jambs.

Remodeling: Framing a Doorway

Relocating kitchen cabinets and appliances, subdividing a room, or simply redirecting traffic flow around a work area may require cutting and framing a new doorway. Before you cut into any wall, be sure that it's nonbearing (see page 69). If the wall contains electrical wires, pipes, or ductwork, they must be rerouted.

Positioning the opening. You'll need to plan an opening large enough for both the rough opening and the rough door framing—an additional 1½ inches on top and sides.

Often it's simplest to remove the wall covering from floor to ceiling between two bordering studs (the new king studs) that will remain in place. In any case, you'll save work later if you can use at least one existing stud as part of the rough framing. Locate the studs in the area; then, using a carpenter's level for a straightedge, draw plumb lines to mark the outline of the opening on the wall.

Removing wall covering and studs. First remove any base molding. If your wall covering is gypsum wallboard, cut along the outline with a reciprocating or compass saw, being careful to sever only the wallboard, not the studs beneath. Pry the wallboard away from the framing. To remove plaster and lath, chisel through the plaster to expose the lath; then cut the lath and pry it loose.

Next, you'll need to cut the studs inside the opening to the height required for the header. (For a partition wall, you can use a single 2 by 4 header laid flat.) Using a combination square, mark these studs on the face and one side, then cut carefully with a reciprocating or crosscut saw. Pry the studs loose from the sole plate.

Framing the opening. With wall covering and studs removed, you're ready to frame the opening. Measure and cut the header, and toenail it to the king studs with 8-penny nails. Nail the header to the bottoms of the cripple studs with 16-penny nails. Cut the sole plate flush with the bordering king studs, and pry it loose from the subfloor.

Now cut trimmer studs; nail them to the king studs with 10-penny nails in a staggered pattern. You'll probably need to adjust the doorway's width by adding spacer blocks and an extra pair of trimmers on one side, as shown below right.

HOW TO CUT & FRAME A DOORWAY

Remove wall coverings, then cut studs within the opening even with the top of the new header.

Nail header and trimmer studs to the king studs; block out an extra pair of trimmers, if needed.

Windows

Never has there been a greater variety of windows from which to choose. The four most popular styles are double-hung, sliding, casement, and fixed.

Double-hung windows, the most widely used of all styles, have two sashes—an upper, outside sash that moves down, and a lower, inside sash that moves up in grooves in the frame. *Sliding* windows have movable sashes that slide in horizontal tracks. *Casement* windows, hung singly or in pairs, have sashes that swing outward; the modern ones are operated with a crank. *Fixed* windows, often called "picture" windows, are stationary units mounted within a frame.

Most movable windows are available as prehung units, a condition that greatly simplifies their installation. Fixed windows can be purchased prehung, or they can be mounted within an existing frame or a frame of your own construction.

You can install any of these windows either in a new wall or in an old wall that you'll need to open up and reframe. For details on framing a new wall, see pages 69–71; if you're remodeling, this section will show you how to open up the wall and reframe the window opening.

Installing a Prehung Window

A prehung window comes complete with sash, frame, sill, hardware, and all the trim except the interior casing. It's easiest to install one of these windows from the outside, though you can maneuver it through the rough opening from the inside and lean through the window to fasten it.

Installation: New wall. Depending on the design you choose, you'll secure prehung windows within the rough opening by either 1) nailing through a flange that surrounds the window; 2) nailing through the exterior trim on the outside of the window; or 3) nailing or screwing the jambs through the shims into the trimmer studs and header surrounding the rough opening. In new construction, windows are normally installed after the sheathing is on, but before the siding is attached. If there's no sheathing on your walls, mount the window unit directly over studs and building paper.

Center the window in the rough opening and have a helper hold it securely against the wall. From inside, raise the window to the correct height above the rough sill with shims or wood blocks. Check the top of the unit with a carpenter's level. If it's not level, adjust the shims.

Depending on your window's installation method, either tack in place the bottom corners of the mounting flange or exterior trim, or drive nails or screws through the finish sill and shims into the rough framing below.

Now plumb and level the head and side jambs with more shims. Check the corners with a carpenter's square.

To fasten a flange-mounted window, space 6-penny galvanized common nails about 8 inches apart. Nail through outside trim or jambs with 8-penny galvanized finishing or casing nails spaced every 16 inches.

Installation: Existing wall. If you're mounting the window in an existing wall, first center the window in the opening; then, using wood shims or blocks, raise it to the desired height above the rough sill. Check the level of the head jamb, then have your helper trace the outline of the flanges or trim. Remove the window. Using a circular saw with the blade set to cut through the siding but *not* through the sheathing or wall framing, cut along the marked lines; remove the siding. Set the window back in the opening, level and plumb it, and fasten it as described above.

Finishing the job. Unless your roof has a pronounced overhang, your building code may require a drip cap or metal flashing over the top of the window. Thoroughly caulk the joints between the siding and the new window. Once interior wall coverings are in place, trim the inside of the window; for techniques, see pages 110–111.

BASIC WINDOW ELEMENTS

A prehung window unit arrives ready to install. Simply position it inside the rough opening, level and plumb the jambs, and fasten the unit to the framing through the jambs or casing.

...Windows

Building a Simple Fixed Window

Some windows, usually enclosed within wood or aluminum sashes, can be installed in existing frames. Or you can simply have a glass dealer cut glass to your frame size and then install it yourself in the frame as a fixed window.

If you don't have a frame, you can use a little woodworking skill and an electric router to build a solid one of your own.

Recycling an existing frame. To reuse a casement window's frame, you'll first need to remove the existing sash, molding, and hardware. With a double-hung window, pry off the inside stop and remove the lower sash and balance. Then pull out the parting strips to remove the upper sash and balance.

If you're planning to repaint any part of the old frame, you'll save work by doing the preparatory work now.

Building a new frame. To build a simple, solid window frame, choose clear, knot-free lumber—pine is a good choice. Consider 1-by lumber (¾ inch thick) the minimum; for a large window, choose 2-by lumber. The lumber width should match the thickness of your wall, including outside sheathing or sheet siding and inside wall coverings.

A window frame is basically a simple box. But the bottom piece, or finish sill, should be sloped (14° is standard) to allow water to run off. You can buy premade sills in wood or metal, or construct your own. If you choose to make your own, cut a thin groove below the outside lip to serve as a "drip edge."

Lay out the dadoes near the tops of the side jambs, as shown below. Cut them with an electric router (for tips, see pages 16–17). Also dado the bottoms of the jambs or angle the jamb ends to match the sill's slope.

Assemble the frame with finishing nails and waterproof glue.

Mounting an aluminum window. A new aluminum window is held in place by stop molding nailed on the inside and outside. Using finishing nails, nail one set of stops to the side and head jambs. Following the manufacturer's instructions, caulk along the stops and frame with a nonhardening glazing compound; this seals the joint around the window. Set the window into position, add another bead of glazing around the perimeter, and carefully nail a second set of stops in place. The new stool (see page 111) may also serve as the bottom stop.

Installing plate glass. Ask the glass dealer for recommendations on glass type and thickness, then have it cut to fit your frame. You may need small glazier's clips, rubber seals or gaskets, or metal or plastic moldings to space the glass away from the stops and frame. The glass should not touch either frame or stops, but should "float" in a layer of glazing compound. If the installation seems difficult, have the dealer mount the glass in the frame for you.

Installing the new unit. At this point, the installation of a fixed window is similar to that of a prehung unit secured through the jambs. Position your unit inside the rough opening; then shim, level, and plumb as described on page 97.

ASSEMBLING A FIXED WINDOW

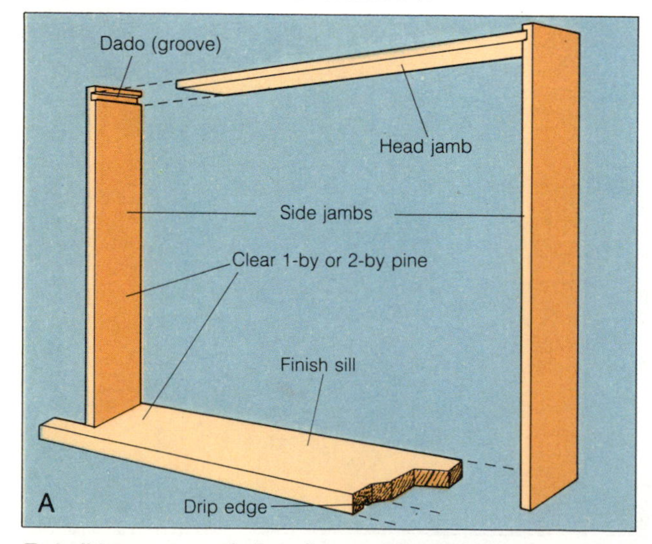

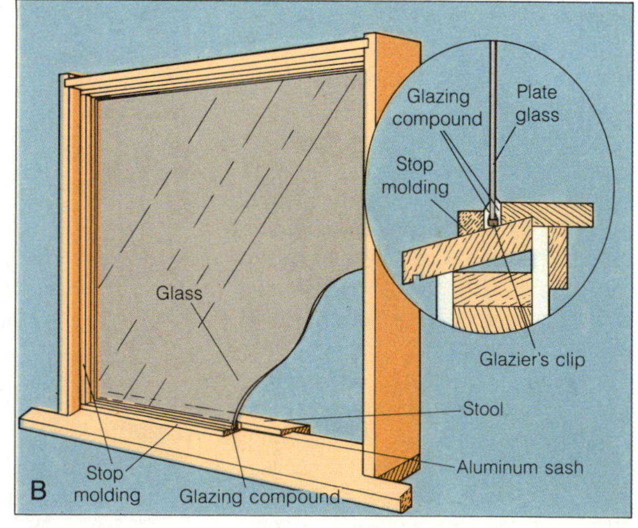

To build your own window, first recycle an old frame or construct your own (A). Install an aluminum-sash window in the completed frame (B), or mount plate glass as shown in the inset at right.

Remodeling: Framing a Window Opening

Cutting and framing a window opening in an existing wall is similar to framing a doorway (see page 96), though in addition you must cut into the exterior siding and sheathing of the house. And you may be dealing with a bearing wall (see page 69) or balloon framing (see page 59); both call for additional steps.

Locating the opening. Before you start, figure out the precise width of the rough opening you need: check the manufacturer's specifications or simply add ⅜ inch to each side (¾ inch total) of the actual width of the window unit you plan to use. Mark this width on the wall where you'd like the window. Now, evaluate the position of the window in relation to the underlying studs. If you can use at least one existing stud as a king stud, you'll save work. You'll need to reroute any pipes or wires that cross the opening, unless you can reposition the window slightly.

Cut through the wall coverings and pry the coverings loose from the wall.

Marking the studs. To lay out the top of your opening (and the top edge of the header), first measure the height of existing doors and windows; then add on the ⅜-inch shimming space and the height of the header. To determine the correct header size for your opening, see the chart on page 70 and check local codes. Unless your code demands something heavier, the header for a nonbearing wall can be a single 2 by 4 or 2 by 6 laid flat.

Using a combination square and pencil, mark the total height on the king studs flanking the opening. Then transfer the height to the studs within the opening. Mark each stud on the edge and one side.

Now locate the bottom edge of the rough opening (and the bottom of the rough sill): from the marks you've just drawn, measure down the height of the rough opening (or the height of the window unit plus ¾ inch), and add 1½ inches more for the thickness of the rough sill. Mark the studs once more. If your wall is a bearing wall or is constructed with balloon framing, *do not* begin cutting until you've supported the ceiling and structure above the opening.

Supporting the structure. To shore up a bearing wall ceiling, you'll need to build a temporary wall slightly longer than the width of the opening and about 4 feet away from the existing wall. Protect the flooring with a piece of ½-inch plywood slightly longer than the 2 by 4 plates. Measure the height from floor to ceiling and cut 2 by 4 studs to this dimension, minus 4 inches. Assemble the wall on the floor, spacing studs on 16-inch centers. (For wall-building details, see pages 69–70.)

Raise the wall into position, then slip towels or old rags between top plate and ceiling. Next, while a helper holds the wall plumb, drive shims between the sole plate and plywood until the plate is tight against the ceiling. Position shims under each stud.

In a balloon-framed house, you'll also need to support the studs above the opening. You do this by placing a 2 by 8 *waler* (a horizontal board) against the ceiling and attaching it to the studs with lag screws. Wedge 4 by 4 posts (or posts made from two 2 by 4s) between the waler and the floor.

Framing the new opening. With a reciprocating saw or handsaw, carefully cut the studs to be removed at the lines marking the bottom of the rough sill and at the header's top edge. These cut studs are the cripple studs.

Measure the distance from the sole plate to the line marking the sill on each king stud; cut two additional cripple studs to this length. Fasten these cripples to the king studs with 10-penny nails. Now measure the distance between king studs; cut the rough sill to this length and nail it to each cripple with two 16-penny nails.

For a bearing wall, you'll need to assemble a header, as discussed on pages 70–71. Have your helper position the top of the header, crown side up, against the upper cripple studs. Using 8-penny nails, toenail the ends of the header to the king studs and nail the cripples to the header.

Finally, measure the distance between the bottom of the header and the top of the sill. Cut two trimmer studs to this length and nail them to the king studs with 10-penny nails. To adjust the width of your opening, block out a doubled trimmer on one side, as shown below, being sure to allow enough shimming space.

Opening up the outside. On a day with zero probability of rain, drill a hole through the wall from inside at each corner of the rough opening. Stick a nail through each hole so you can find the corners outside.

From outside, mark the outline of the rough opening with a pencil and a level. Then, using a reciprocating or circular saw, cut through the siding and sheathing along the lines and lay the cut material aside.

REFRAMING THE OPENING

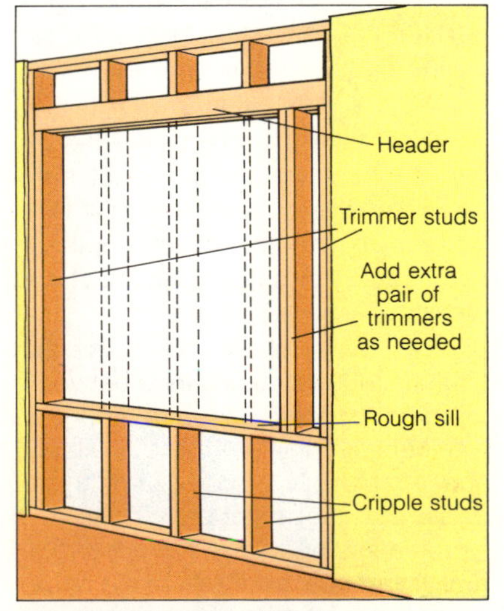

Framing the new opening means cutting away existing studs and adding cripples, rough sill, header, and trimmers.

Basic Stairways

Converting an attic or loft, finishing a basement, or adding a deck is likely to mean that you must consider stairs. The types of stairways described here are "rough" (that is, suitable for outdoor, basement, and attic applications). They're easier to build than they look, provided you follow certain basic rules, measure carefully, and plan each detail before you begin construction. Should you need to frame a new opening, or *well*, in the floor or ceiling for your stairway, see page 67 for details.

Major stairways serving interior living spaces can be a great deal more complicated; these often involve decorative woodwork or changes in direction and elevation that require considerable skill in planning and execution. Such stairs go beyond the scope of basic carpentry, but the principles are the same as those discussed below.

Planning: The Key to Stair-building

Three calculations are critical to your stairway's plan: tread depth, riser height, and the stairway's angle. You must also consider stair width, height of railing, and headroom.

Figuring your layout. The angle of a stairway is a function of its riser-tread relationship. If the angle is too steep, the stairs will be a strain to climb. The ideal angle lies between 30° and 35°.

Normally, the sum of the riser height and tread depth should be 17 to 18 inches. Ideal riser height is 7 inches (many building codes specify a maximum of 7½ inches).

To determine the number of steps you'll need, measure the vertical distance—in inches—from finish floor to finish floor and divide that measurement by the ideal riser height, 7 inches. If the answer ends in a fraction (as it probably will), drop the fraction and divide the whole number into the vertical distance; the resulting figure will give you the exact measurement for each of your risers.

Now take the exact riser height and subtract it from the ideal sum for both risers and treads—17½ inches—to find the exact depth of each tread. For safety, they must not measure less than 10 inches.

Next, you need to determine the total run (horizontal distance between top and bottom risers) to find out whether your plan will fit the available space. Multiply your exact tread depth by the number of risers—minus one—and you'll have the total run. If this run won't fit your space, simply adjust the riser-tread relationship, increasing one and decreasing the other, until you achieve a total run that will work. If a straight run proves too long, you may have to change to a U (180° change in direction) or L (90°) design.

Headroom, width, railings. To avoid hitting your head or having to duck every time you ascend or descend interior stairs, you must provide adequate headroom. Most codes require a minimum headroom (measured from the front edge of the tread to any overhead obstruction) of 6½ feet. About 7½ feet is ideal to avoid giving the stairway a closed-in feeling.

The width of the stair is less important and will be dictated largely by the space available, but it should allow two persons to pass. Building codes usually specify a minimum of 30 inches; a width of 36 to 42 inches is preferable.

Handrailings 30 to 36 inches high (measured from the top of the tread to the top of the railing) are comfortable for a person of average stature; 34 inches is a good height above floors and landings.

Choose your materials. Because it provides the greatest strength, the best stringer design is the single-piece stringer with "sawtooth" cutouts for the steps. You'll need knot-free 2 by 12 dimension lumber long enough to reach from the top landing to the bottom flooring.

For risers and treads, select a knot-free grade of lumber not less than 1 inch thick (nominal dimension). It's common to use 2-inch boards for treads and 1-inch for risers. You may also want to purchase precut treads and risers.

STAIR-BUILDING TERMS ILLUSTRATED

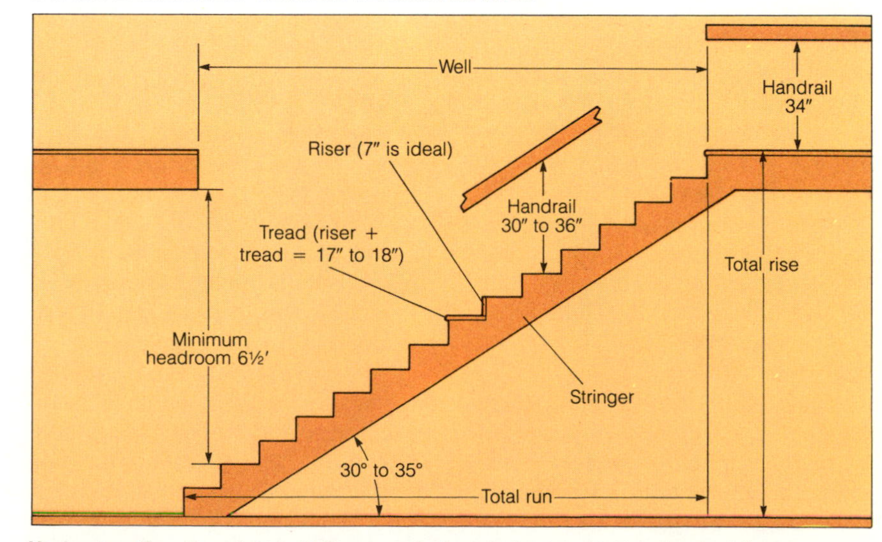

Understanding the stair-building vocabulary is the key to planning a stairway.

Building the Stairway

The typical stair assembly consists of stringers, risers, treads, and railings. Here's how to assemble each element in turn.

Stringers. The procedure for laying out a sawtooth stringer is similar to that for stepping off a rafter (discussed on page 77). To lay out the cuts, mark the dimension of the risers on the tongue of a carpenter's square; then mark the dimension of the treads on the square's body. Line up the marks with the top edge of the 2 by 12 stringer, as shown at right, and trace the outline of the risers and treads onto it.

Cut out the notches with a handsaw or circular saw, finishing with the handsaw. Because the tread thickness will add to the first step height, measure the exact thickness of a tread and cut this amount off the bottom of the stringer. Once your pattern is cut, check the alignment; then, if it's satisfactory, use it as a pattern to mark the second stringer. If your stairway is 36 inches or wider, you should plan to space a third stringer midway between the end stringers.

Generally, nailing the top of the stringer to the rough opening's trimmers or headers (see page 67 for details) is sufficient, but you may want to increase strength by adding an extra header board, metal joist hangers, or plywood ledger (see drawing at right). At the bottom, either toenail the stringers to the floor or notch them for a 2 by 4 ledger, as shown.

If one or both end stringers will be "closed"—that is, attached to an adjacent wall—first nail an additional 1 by 12 plate to the wall studs: this acts both as trim and as a nailing surface for the main stair stringer.

Risers and treads. When measuring and cutting risers and treads, remember that the bottom edge of the riser overlaps the back of the tread, and the forward edge of the tread overlaps the riser below it. Giving each tread a 1¹/₈-inch *nosing* (a projection beyond the front of the riser) lends a more finished appearance to the stairway.

Nail risers to the stringers first, using 8-penny finishing nails. Then nail the treads to the stringers with 12-penny nails. Finally, fasten the bottom edges of risers to the backs of treads with 8-penny nails. Gluing treads and risers to the stringer as you nail them will help minimize squeaks.

Railings. Whether you use a simple length of 2 by 4 or purchase finished decorative railing, fasten it securely to an inside wall by screwing or bolting commercial brackets to wall studs (every third stud). For the open sides of stairways, begin with sturdy posts not less than 2 inches square; bolt them directly to the stringer. Cap rails for outdoor and rough stairways are usually 2 by 4s or 2 by 6s nailed to the top of each supporting post.

LAYING OUT A STRINGER

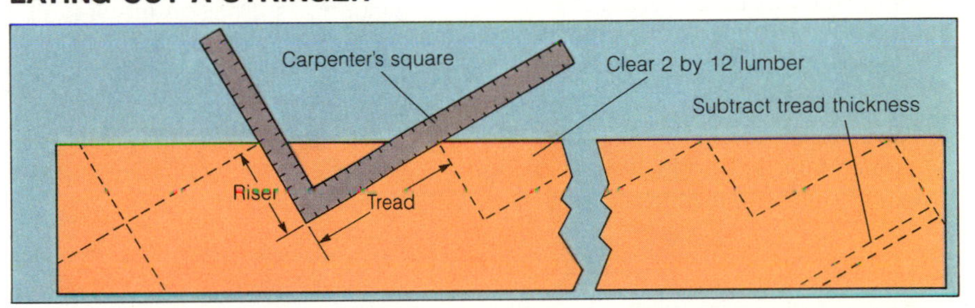

To mark a stringer for cutting, align the square's marks that correspond to riser height and tread depth with the 2 by 12's edge; then trace along the square.

THE FINISHED STAIRWAY

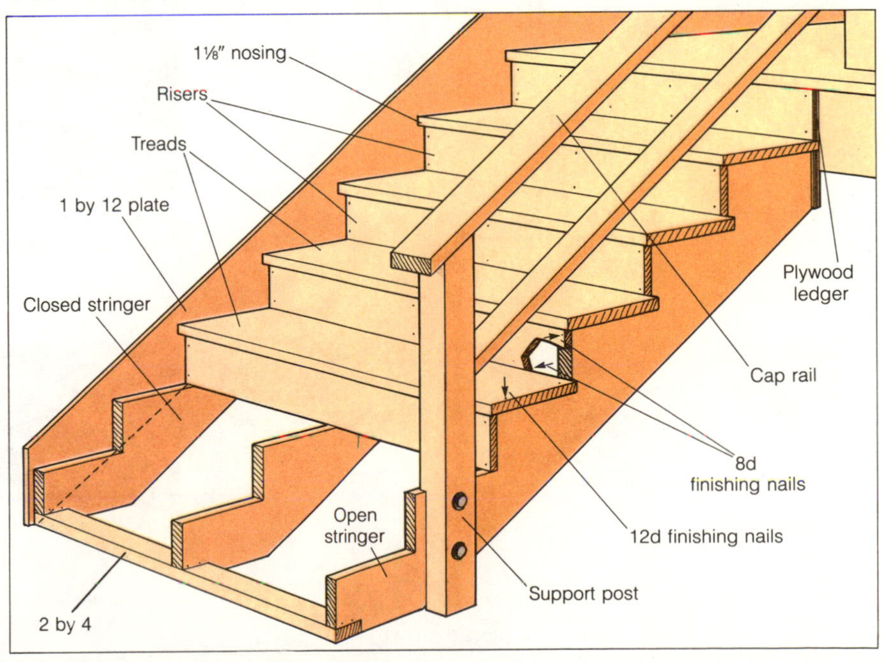

The "rough" stairway consists basically of stringers, risers, and treads; support posts and rails complete the picture.

Ceilings

In most homes, the ceilings are not the outstanding feature. Even so, a new ceiling can brighten up a room with light, color, and a clean, finished appearance.

Suspended panel ceilings and those consisting of mineral and wood-fiber tiles are easily installed. Gypsum wallboard, today's standard, is not technically complicated to hang, but the large, heavy panels are awkward to handle. These three ceiling types are discussed in detail below.

Besides wallboard, a host of wall coverings—solid board and plywood panelings and even siding—are suitable for ceilings. To apply these materials, simply adapt the directions found in "Siding" (pages 82–87) and "Interior Wall Coverings & Trim" (pages 106–111).

Putting Up Ceiling Tiles

Square and rectangular ceiling tiles are available in several decorative and acoustic styles. These tiles, most commonly 1 foot square, can be applied either directly to existing, flat ceilings in good shape or to 1 by 3 furring strips fastened across joists or ceiling with special adhesive or staples.

To estimate the number of tiles you'll need, follow the method described for suspended ceilings on the facing page. Plan to trim border tiles to equal widths along opposite edges of the room.

New construction. To form a solid nailing base for tiles, fasten furring strips to ceiling joists with 6-penny common nails. Position the first strip along the edge of one wall, perpendicular to joists. Place the second strip so that the edges of the border tiles will be centered on the strip. Then space each succeeding strip 12 inches on center.

Beginning in one corner, cut and install border tiles. Place the cut edges against the wall and nail them to the furring. Then staple these and all remaining tiles through the flanges.

Tiling over a ceiling. If you're applying tiles over an existing ceiling, first mark your layout across the ceiling by snapping a chalkline for each row. Install the tiles by daubing special adhesive on each corner and the center of each tile's back.

A TILE CEILING: FOUR STEPS

A grid supports ceiling tiles. Nail the first furring strip flush against the wall at a right angle to joists.

Level the furring strips by driving shingles between strips and joists. Strips must also be level with each other.

Staple border tiles through flanges after cutting them to size. Face-nail to furring strips where molding will cover nailheads.

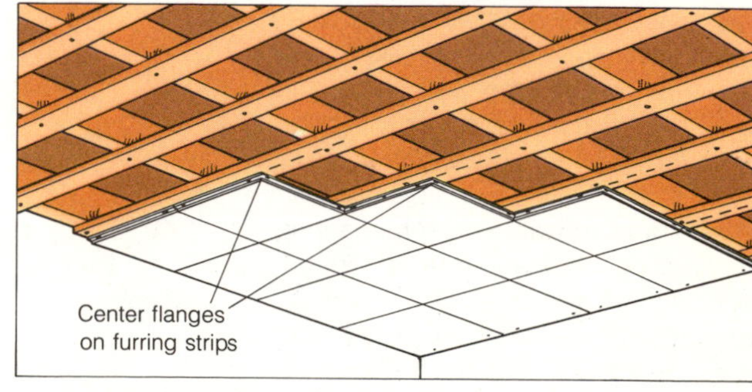

Work outward from border tiles across the room, centering each tile on furring strips.

Hanging a Suspended Ceiling

Easy-to-install suspended ceilings consist of a metal grid supported from above by wire or spring-type hangers. The grid holds acoustic or decorative fiberboard panels.

The most common panel size is 2 feet by 4 feet, though panels are available in a variety of sizes. Transparent and translucent plastic panels and egg-crate grilles are made to fit the gridwork and admit light from above. Recessed lighting panels that exactly replace one panel are also available from some manufacturers. All components are replaceable, and the panels can be raised for access to the area above.

Figuring your needs. To determine the number of panels you'll need, first measure your wall lengths at the proposed ceiling height. Draw the ceiling area to scale on graph paper, using one square per foot of ceiling size. Block in the panel size you'll be using. Finally, count the blocked areas and parts of areas to get the number of panels you'll need.

For a professional-looking job, plan equal borders on opposite sides of the room. To determine the nonstandard width of panels needed for perimeter rows, measure the extra space from the last full row of panels to one wall, and divide by two. This final figure will be the dimension of border panels against that wall and the opposite wall. To complete your plan, repeat this procedure for the other room dimensions.

Installing the ceiling. First, figure the ceiling height—at least 3 inches below plumbing, 5 inches below lights; minimum ceiling height is 7 feet 6 inches. Snap a chalkline around the room at your chosen level and install right-angle molding with its base on the chalkline.

Next, install the main runners as shown below. Cut the runners to length with tinsnips or a hacksaw. Setting them on the right-angle molding at each end, support them every 4 feet with #12 wire attached to small eyescrews fastened into joists above. Lock 4-foot cross tees to the main runners by bending the tabs in the runner slots.

Set the panels into place and install any recessed lighting panels. Cut border panels as necessary with a sharp utility knife. Be sure your hands are clean when handling panels—smudges and fingerprints are hard to remove.

THE SUSPENDED CEILING SEQUENCE

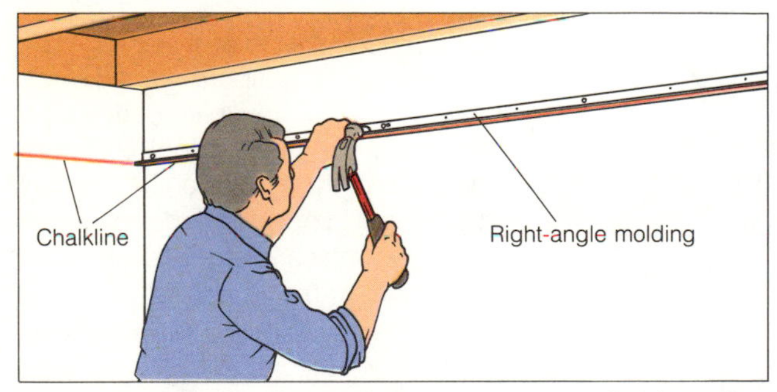

First, snap a chalkline around the room and install right-angle molding with its base on the chalkline.

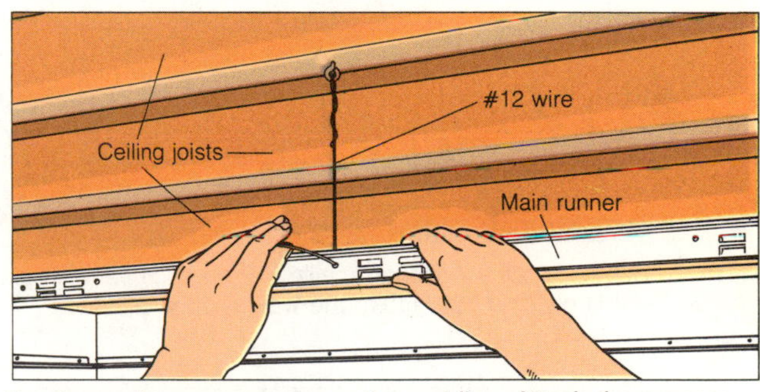

Set the main runner ends on the molding. Attach the runners to the ceiling joists with #12 wire.

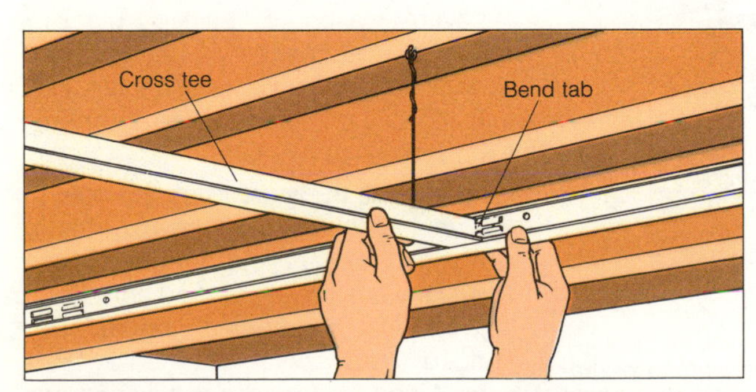

Lock 4-foot cross tees to the main runners by inserting the tabs through the runner slots and then bending the tabs.

Install the panels by sliding them up diagonally through the grid openings and then lowering them into place.

. . . Ceilings

Gypsum Wallboard for Ceilings

Though gypsum wallboard panels are heavy and awkward to install on ceilings, they're the most popular choice because they are inexpensive, they take paint and surface textures well, and one panel covers a large area. A few special techniques will help the work go smoothly—and will lead to a smooth result, as well.

Basic ceiling application. Methods for installing a wallboard ceiling are basically the same as those for walls (see pages 106–107). Choose ½ or ⅝-inch-thick panels, and fasten them perpendicular to joists with annular ring nails, drywall screws, or a combination of nails and construction adhesive. Nail spacings are governed by local codes, but typical spacing is every 7 inches along panel ends and at intermediate joists (called "in the field").

If you decide to double-nail (see page 106)—a smart choice for ceilings—space the first set of nails every 7 inches along the ends and every 12 inches in the field. Then place a second nail about 2 inches away from each nail in the first set. Nails should be spaced at least ⅜ inch in from the edges around the perimeter.

If your walls are also being finished with wallboard, be sure to apply the ceiling first; the edges of ceiling panels may be supported by wall panels.

Tips for a safe, smooth job. Because it's necessary to support the heavy panels while fastening them, installing a wallboard ceiling is a two-person job. First, position a pair of stepladders, or set up a couple of sturdy sawhorses, laying a few planks across them to serve as a short scaffold to stand on. Then, both carpenters hold their respective ends of a panel in place with their heads. Begin nailing at the center of each panel; then place the next few nails where they will take the weight off your heads.

As an alternative, you can construct one or two T-braces as shown below. The length of the braces should equal the height from the floor to the ceiling joists; when the panel is positioned, the extra thickness will help wedge the brace in place.

Supporting the panel ends. The long edges of wallboard panels are tapered to allow you to smoothly tape and finish the joints. Panel ends are not tapered, though; when two panels butt, this joint is hard to finish smoothly unless the joists are perfectly level with one another.

Many professionals solve this problem by "floating" the end joint between adjacent joists. Install the first panel with the end roughly halfway between joists, and nail it to the last joist before the gap. Now, cut several wallboard "backing blocks," about 8 inches by 12 inches, and spread joint compound on the bottoms. Lay the blocks atop the panel end, as shown at left.

Install the second panel, butting it against the first. Push a 1 by 4 strip against the joint and hold it in place with several braces tacked across the joists. After the backing blocks have dried, the depression formed by the 1 by 4 can be taped and finished just like the tapered edges.

TIPS FOR HANGING WALLBOARD CEILINGS

It takes two to install a wallboard ceiling. Prop each panel in place with your heads; space the first nails so they'll take the weight off your heads.

T-braces, used singly or in pairs, will also hold panels in position. Fashion a simple, sturdy brace from a 2 by 4 upright and a 1 by 4 crosspiece.

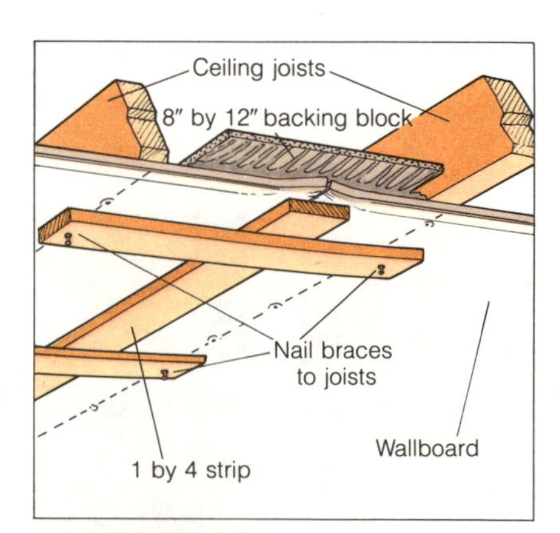

Backing blocks allow you to "float" panel ends between joists for an extra-smooth joint. Braces and a 1 by 4 hold panels tight while joint compound dries.

REMODELING WITH SKYLIGHTS

Installing a skylight in an existing pitched roof is a three-part process: you cut and frame openings in both roof and ceiling, mount the skylight unit, and then connect the two openings with a vertical, angled, or splayed light shaft. A brief outline of the skylight installation sequence follows; for a complete discussion of techniques, see the *Sunset* book *Windows & Skylights*.

Marking the openings. After planning the layout of your roof opening, light shaft, and ceiling opening, mark the location of the ceiling opening; then drive nails up through the four corners and center so they'll be visible in the attic or crawlspace. From the attic, check for obstructions, shifting the location if necessary.

Use a plumb bob for transferring the ceiling marks to the underside of the roof; again, drive nails up through the roofing materials to mark the location.

Framing the roof opening. On a day with zero probability of rain, cut and frame the roof opening.

When you work with a skylight designed to be mounted on a curb frame, build the curb first; 2 by 6 lumber is commonly used. Your skylight may have an integral curb or may be self-flashing; if so, you can skip this step.

To determine the actual size of the opening you need to cut, add the dimensions of any framing materials (see below) to the rough opening size marked by the nails. You may need to remove some extra shingles or roofing materials down to the sheathing to accommodate the flashing of a curb-mounted unit or the flange of a self-flashing unit.

Cut the roof opening in successive layers: roofing materials first, sheathing next, and finally any necessary rafters. Before cutting the rafters, support them with 2 by 4s nailed to the ceiling joists below.

To frame the opening, you'll need double headers and possibly trimmers. Install the headers with double joist hangers.

If you're installing a curb-mounted unit, position and flash the curb. Toenail the curb to the rafters or trimmers and to the headers. Pay special attention to the manufacturer's instructions concerning flashing.

Mounting the skylight. For a curb-mounted unit, secure the skylight to the top of the curb with nails and a sealant. Set a self-flashing unit in roofing cement and then nail through the flange directly to the roof sheathing. Coat the joints and nail holes with more roofing cement.

Opening the ceiling. Double-check your original ceiling marks against the roof opening and the intended angle of the light shaft. Cut through the ceiling materials and then sever the joists. Support joists to be cut—do this by bracing them against adjacent joists. Frame the opening in the same manner as for the roof opening.

Building a light shaft. Measure the distance between the ceiling headers and roof headers at each corner and at 16-inch intervals between the corners. Cut studs to fit the measurements, and install them as illustrated above. This provides a nailing surface for wall coverings.

Final touches. Insulate the spaces between studs in the light shaft before fastening wallboard to the studs. To maximize reflected light, paint the wallboard white.

Trim the ceiling opening with molding strips. Adding a plastic ceiling panel (either manufactured or cut to size) helps diffuse light evenly.

BASIC SKYLIGHT COMPONENTS

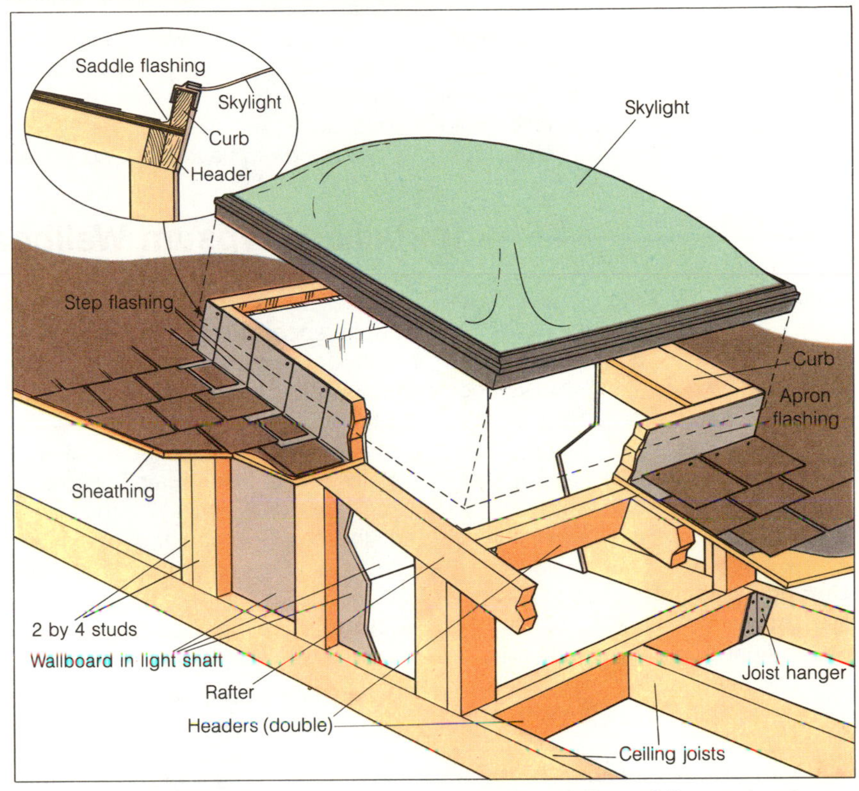

When adding a skylight, you'll reframe roof and ceiling, then carefully mount and flash the unit of your choice.

Interior Wall Coverings & Trim

Walls and trim are prominent features of almost any interior. Carefully applied, your wall coverings—whether gypsum wallboard, wood sheet paneling, or solid boards—can be a real asset. Moldings, whose function is to hide the raw edges of wall coverings, also introduce new architectural interest to a room.

You can attach wall coverings to new studs, directly over existing wall coverings, or to furring strips fastened to the wall.

Arrange to have all wood paneling and trim stored for at least 2 days in the room where it will be installed—the ideal storage time would be a week to 10 days. This allows the material to adapt to the room's temperature and humidity, preventing later warping or buckling.

Installing Gypsum Wallboard

Cutting and installing gypsum wallboard is a straightforward job, but concealing the joints between panels and in the corners demands patience and care. And the weight of full panels can be awkward to negotiate. Wallboard is easily damaged; you'll need to take care not to bend or break the corners or tear the paper covers.

Cutting wallboard. To make a straight cut, first mark the cutting line on the front paper layer with a pencil and straightedge, or snap a chalkline. Cut through the front paper with a utility knife.

Turn the wallboard over and break the gypsum core by bending it toward the back. Finally, cut the back paper along the bend, and smooth the edge of the cut with a perforated rasp.

When fitting wallboard around obstructions such as doorways or windows, carefully measure from the edge of an adjacent wallboard panel or reference point to the obstruction. Transfer the measurements to a new panel and make the necessary cuts with a wallboard or compass saw. For small cutouts like electrical outlet or switch boxes, first rub a bit of colored chalk (or even lipstick) on the edges of the box. Position the panel and press in on the area; then saw along the outline left by the chalk. If the fit is too tight, trim with a perforated rasp.

Basic installation. Wallboard panels may be positioned either vertically or horizontally—that is, with the long edges either parallel or perpendicular to wall studs. Most professionals prefer the latter method because it helps bridge irregularities between studs and results in a stronger wall. *Don't* use this method, though, if your wall is higher than 8 feet—the extra row requires more cutting and creates too many joints.

INSTALLING THE PANELS

Lift each wallboard panel into position and center the edges over wall studs. Then nail the panel to the studs, dimpling the wallboard surface slightly with the hammer.

Before installing panels, mark the stud locations on the floor and ceiling. Starting from one corner, lay the first panel tight against the ceiling. If you choose the horizontal method, panel ends may be either centered over studs or "floated" and tied together with backing blocks, as discussed on page 104. Stagger the end joints in the bottom row so they don't line up with the joints in the top row.

Fasteners. Wallboard may be fastened to walls with nails, drywall screws, or construction adhesive supplemented by nails. Fastener spacings are subject to local codes, but typical nail spacing is every 8 inches along panel ends and edges and along intermediate supports ("in the field"). Be sure that nails are no closer to the edges than 3/8 inch.

Panels can also be "double-nailed," ensuring that if one nail fails, its partner will hold the panel. When double-nailing, space a second nail 2 inches from each nail in the first set; you can space these pairs 12 inches apart in the field.

For best results when nailing, use ring-shank nails with 1/4-inch-diameter heads. Choose 1 1/4-inch nails for 3/8 and 1/2-inch panels, or 1 5/8-inch nails for 5/8-inch panels.

Drive in the nails with a bell-faced hammer. Your goal is to dimple the wallboard surface without puncturing the paper. If you do puncture the paper or miss a stud, pull out the nail and install another one. The hole can be patched and sanded later. It's usually simplest to first tack a row of panels in place with a few nails through each; later you can snap chalklines to mark the studs, and then finish the nailing pattern.

TAPING WALLBOARD JOINTS

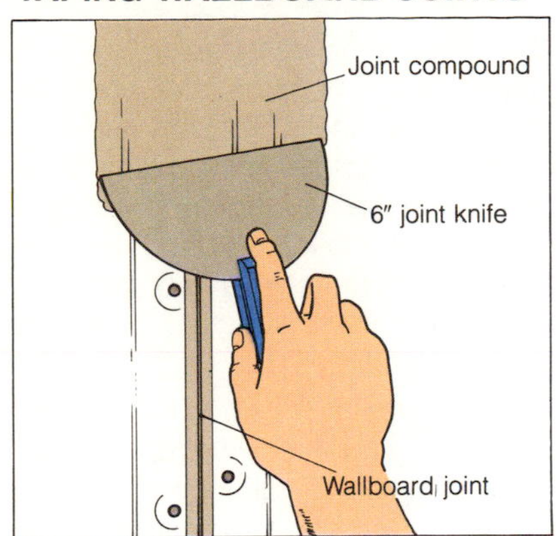

First, spread a smooth layer of joint compound over the joint with a 5″ or 6″ joint knife.

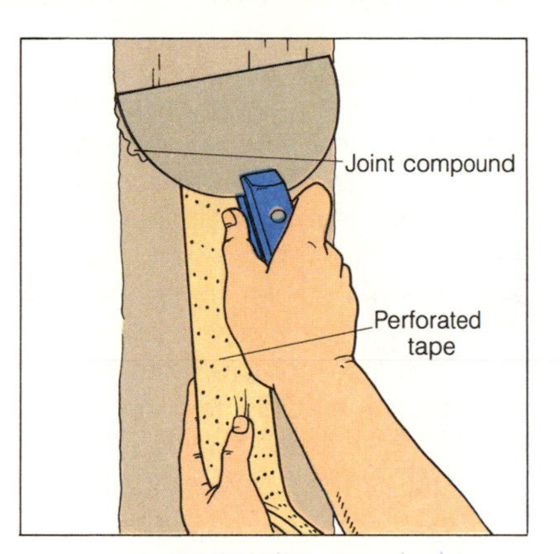

Embed perforated fiber tape in the compound, and apply a second, thinner layer of compound on top.

After the compound is dry and sanded, apply a wider layer with a 10″ or wider taping knife, feathering the edges.

If your wallboard will serve as a backing for ceramic tile, paneling, or cabinets, you may not need to hide joints and corners. But if you expect to paint or wallpaper, you'll want to finish the wallboard.

Taping joints and corners. To finish wallboard neatly, you'll need wallboard tape (buy tape that's precreased) and joint compound. Premixed joint compound is much simpler to use than the powdered variety. Both kinds can contain asbestos; compounds that do not are marked on the label.

The taping process has three stages. To tape a joint between panels, first apply a smooth layer of taping compound over the joint with a 6-inch taping knife. Before the compound dries, embed wallboard tape into it and apply another thin coat of compound over the tape, smoothing it gently with the knife. Use only enough compound to fill the joint and cover the tape evenly; excess compound just means more sanding later.

To tape an inside corner, apply a smooth layer of compound to the wallboard on each side of the corner. Measure and tear the tape to length, fold it in half vertically along the crease, and press it into the corner with a corner tool. Apply a thin layer of compound over the tape and smooth it out with the corner tool.

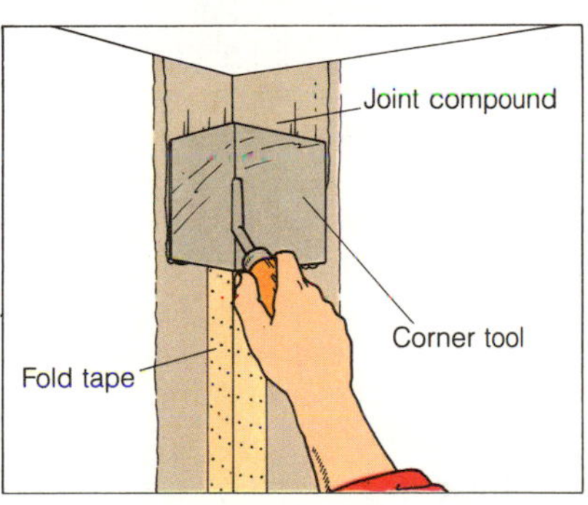

To tape an inside corner, fold tape in half and press it into the compound with a corner tool.

Cover exterior corners with a protective metal cornerbead cut to length and nailed through its perforations. You don't need to tape here: simply run your knife down the sharp metal edge to fill the spaces with joint compound.

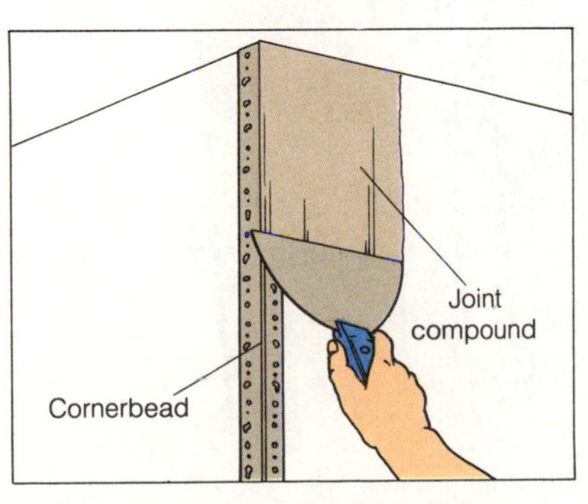

Protect outside corners with metal cornerbead; finish with joint compound.

Continue taping all the joints. Then, using smooth, even strokes with the 6-inch knife, cover the nail dimples in the field with compound.

Allow the taping compound to dry for at least 24 hours before sanding lightly to get a smooth surface. Wear a painter's mask and hat against the dust, and close off the room where you're working.

Now, with a 10-inch knife, apply a second coat of compound, feathering out the edges past each side of the taped joint.

Let the second coat dry. Then sand it and apply a final coat, using the 10-inch or an even wider "finishing trowel" to smooth out and feather the edges. After the compound dries, sand it once more with fine sandpaper to remove even minor imperfections.

Textured versus smooth finish. Although many people prefer the smooth look, texture can hide a less-than-perfect taping job. Also, as the house shifts and framing shrinks over time, any cracks between panels will be less apparent.

Texturing compounds are available ready mixed or in powdered form. Depending on the texture desired, compounds are applied with a brush, roller, trowel, or spray gun. Ask your dealer for recommendations.

. . . Interior Wall Coverings & Trim

Installing Sheet Paneling

Sheet paneling—typically decorative plywood, hardboard, or plastic laminate—can be laid over new stud walls, applied directly to existing walls in good condition, or attached to furring strips over old, bumpy walls. If you're paneling over bare studs, ask your dealer whether you'll need to back the sheets with gypsum wallboard or other material for rigidity and fire protection.

Furring and shimming. Furring strips, usually 1 by 3s or 1 by 4s, are attached to the wall with nails long enough to penetrate studs at least an inch. For masonry walls, use concrete nails or expansion bolts.

Of course, furring strips should be plumb and flat; you can make small adjustments with shingle shims. If the existing wall is severely out of plumb, you may need to block out furring strips at one end.

The correct spacing of furring strips depends on the type of paneling that will cover them. A typical arrangement is shown below. Check the manufacturer's instructions for recommended nail spacings. Be sure to leave a ¼-inch space at both the top and bottom of the wall when applying the strips.

Plan your layout. For a good appearance, cut the first and last panels on a wall the same width unless you're using panels with random-width grooves. Prop up all the panels along the wall to see how they'll fit. Whenever possible, center between-panel joints over door and window openings. At inside corners, plan to butt panels together. Outside corners, unless perfectly mitered, will require two pieces of trim or a corner guard.

Attaching sheet paneling. Adhesives and nails are the two basic fasteners used with most sheet paneling. Adhesive is preferable—it's fast and clean, and it subjects panels to much less risk of dents or nail holes.

Here's the typical procedure: First cut a panel ¼ inch short of the distance from floor to ceiling. On furring or exposed wall framing, apply adhesive to the framing in squiggly lines. On a finished wall, apply adhesive directly to the wall, spacing the squiggly stripes a uniform 12 or 16 inches apart.

Drive four finishing nails through the top edge of the panel—4-penny nails for ¼-inch panels, 6-penny for panels up to ⅝ inch thick, and 8-penny for thicker materials. Position the panel on the wall, leaving a ¼-inch space at the bottom; drive the nails partway into the wall to act as hinge pins. Pull the panel's bottom edge about 6 inches out from the wall and push a block behind to hold it there; wait for the adhesive to become tacky (check the manufacturer's directions for average time).

Then remove the block and press the panel firmly into place. To force the adhesive into tight contact, knock on the panel with a rubber mallet or hammer against a padded block.

Drive the top-edge nails all the way in; then nail the panels at the bottom (you'll eventually cover the nail heads and the ¼-inch gap with molding). Thin paneling materials require either glue or nails within ½ inch of the panel edges to prevent curling.

Scribing a panel. The first piece of paneling that you fit into a corner probably won't exactly match the contours of the adjoining wall or obstruction. To duplicate the irregularities of the adjoining surface on the paneling's edge, first prop the panel into place about an inch from the uneven surface; use shims, if necessary, to adjust level and plumb. Draw the points of a compass or wing dividers along the irregular surface so the pencil leg duplicates the unevenness onto the paneling.

Cut the paneling along the scribed line with a coping saw, saber saw, or block plane.

PUTTING UP SHEET PANELING

To install a panel, first apply adhesive to furring strips or wall, then prop the panel into position as shown. When the adhesive becomes tacky, press the panel into place and finish nailing.

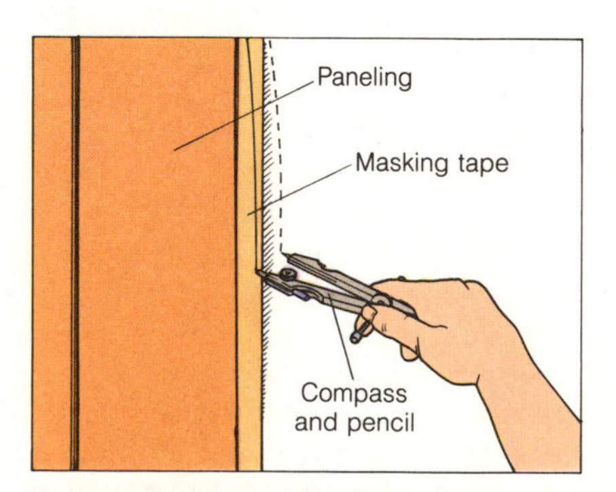

To fit a panel to an irregular surface, scribe along the panel edge with compass or wing dividers.

Cutting an opening. Fitting a panel around any opening requires careful measuring, marking, and cutting.

Keep track of all the measurements by sketching them on a piece of paper. Starting from the corner of the wall or the edge of the nearest panel, measure to the edge of the opening or electrical box; then, from the same point, measure to the opening's opposite edge. Next, measure the distance from the floor to the opening's bottom edge and from the floor to the opening's top edge. (Remember that you'll install the paneling ¼ inch above the floor.)

Marking the side of the panel that will face you as you cut (face up for a handsaw, face down for a power saw), transfer these measurements to the panel. When marking the back of the panel, *remember that measurements will be a mirror image of the opening*.

Solid Board Paneling

Like sheet paneling, solid boards can be attached to new stud walls, to existing walls in good shape, or to a gridwork of furring strips over old, bumpy walls.

Though solid boards are usually installed vertically or horizontally, you don't need to limit your options. For added visual punch, consider using a diagonal pattern or even a pattern designed for exterior siding boards (see page 84 for ideas).

Application methods. You can either nail solid boards to your wall surface or attach them with adhesive. Nailing is the preferred method. For standard 1-by boards, use 6-penny finishing nails and recess the heads 1⁄32 inch below the surface with a nailset. Cover the nail heads, using a putty stick in a matching color.

Vertical pattern. Before paneling vertically with solid boards, you must attach horizontal furring strips every 24 inches on center, or install nailing blocks at these spacings between studs.

Measure the width of the boards you're using and then the width of the wall. From these figures, calculate the width of the final board. To avoid a sliver-size board, split the difference so the first and last boards are the same. Plan to cut boards ¼ inch shorter than the height from floor to ceiling.

When you place the first board into the corner, check the outer edge with a carpenter's level. If the board isn't plumb or doesn't fit the corner exactly, scribe and trim the edge facing the corner (see facing page).

Attach the first board, leaving a ¼-inch space above the floor; then butt the second board against its edge and check for plumb before you nail it. Repeat this procedure with all subsequent boards. To make the last board fit easily into place, cut its edge at a slight angle (about 5°) toward the board's back edge.

At inside corners, simply butt adjacent board edges together, scribing if necessary. At outside corners, you can either miter the joints for a neat fit (cut the miters at an angle slightly greater than 45° so they'll fit snugly) or butt boards and conceal the joints with trim.

Horizontal pattern. Generally, you won't need to apply furring unless the wall is badly damaged or out of plumb. You can nail the boards to the studs directly or through existing wall coverings. To avoid ending with a very narrow board at the ceiling, calculate and adjust its size as described under "Vertical pattern" (preceding).

Start at the bottom of the wall and work toward the ceiling. Nail the first board temporarily at one end, ¼ inch above the floor. Then level the board and complete the nailing. If you need to scribe and trim the board at its ends, follow the instructions on the facing page. Minor inconsistencies can be covered with trim.

Working toward the ceiling, attach each board in the same way. Rip the last board to width as required, leaving ¼ inch of space below the ceiling. Again, if you have trouble fitting the last board, bevel its back edge slightly and pivot it into place.

SOLID BOARD APPLICATIONS

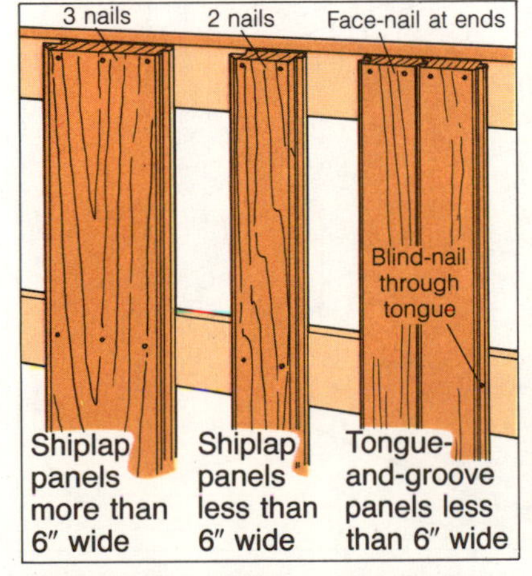

Nailing is the preferred method for fastening solid boards; panel type and size determine exact nail placement.

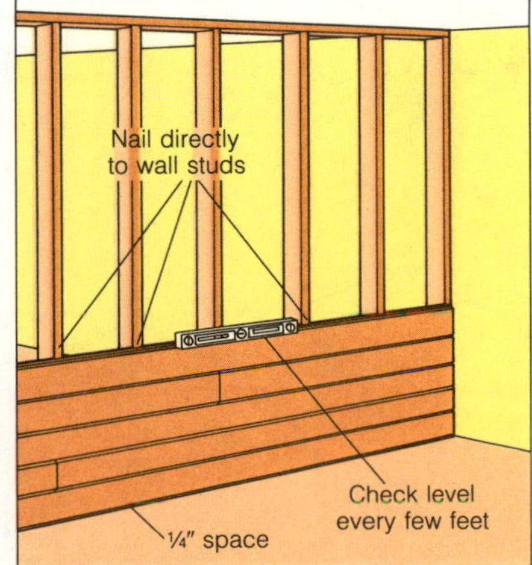

Horizontal patterns can normally be nailed directly to studs; to keep adjustments minor, check level frequently.

...Interior Wall Coverings & Trim

Working with Moldings & Trim

Contoured moldings or standard lumber trim along the bottom edge of wallboard or wood paneling cover the gaps between wall covering and floor. You may also wish to trim the ceiling line and corners. And the edges of both door and window openings need casing. Here are tips to help you create a neat, finished look.

Cutting moldings. A miter box and backsaw are most commonly used for neatly cutting trim. With a miter box, you can cut the precise 45° and 90° angles necessary for most joints. If you're doing a lot of cutting or are working with unusual angles, it may be worth your while to rent or borrow a power miter saw.

Inside joints between contoured moldings placed along baseboards and ceilings must be cut to the proper curvature with a coping saw. For details, see "Baseboards," below.

Fastening moldings. To attach molding, you have three choices: nail it in place with finishing nails and recess the heads with a nailset; fasten it with color-matched nails; or blind-nail it.

To blind-nail, use a small knife or gouge to raise a sliver of wood that's large enough to hide the head of a finishing nail; don't break off the sliver. Pull the sliver to the side, nail into the cavity, and then glue the sliver into place. You can tape the sliver down with masking tape until the glue dries. Rubbing the spot lightly with fine sandpaper will remove all signs of fastening.

Baseboards. Once the finish floor is installed, you can attach baseboards and, if desired, a base shoe (see page 41). These moldings should be installed with a slight gap between the flooring and the bottom of the molding; use thin cardboard as a spacer. Nail the moldings to the wall studs and sole plate, *not* to the floor.

Where two lengths of molding join along a wall, miter the ends to create a *scarf joint.* Nail through the joint to secure the pieces.

Contoured moldings will require a *coped joint* at inside corners for a smooth fit. To form a coped joint, cut the first piece of molding square and butt it into the corner. Then cut the end of the second piece back at a 45° angle, as shown. Next, using a coping saw, follow the exposed curvature of the molding's front edge while reinstating the 90° angle. With a little practice, you can make the contoured end smoothly match the contours of the first piece.

At outside corners, simply cut matching miters in each piece, as shown.

Door casing. Door trim may be either contoured molding or standard lumber. If you choose lumber, plan to butt joints together at the top; for molding, you'll have to miter the joints.

Before installing the casing, pencil a *reveal* or setback line ¼ inch in from the inside edge of each door jamb. Aligning the top casing with the pencil line, mark it where it intersects the side reveal lines. Miter the ends from these points or, if you prefer, add the width of the side casings and cut the ends square. Use 4-penny or 6-penny finishing nails to attach the casing to the jamb, and 8-penny nails along the rough framing. Space nails every 16 inches.

BASEBOARD & CEILING TRIM

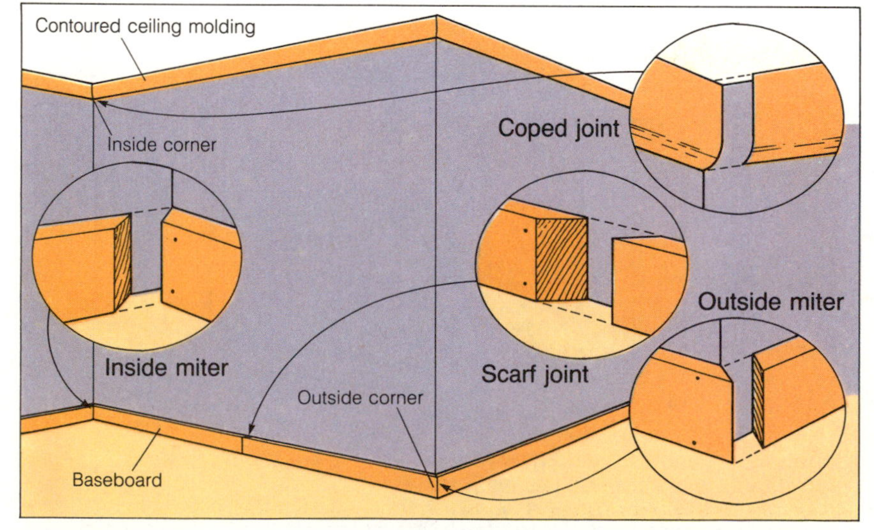

These joints will handle most needs when adding ceiling and baseboard finish. Flat trim calls for basic miter cuts; contoured moldings may require coped joints at inside corners.

COPING A JOINT

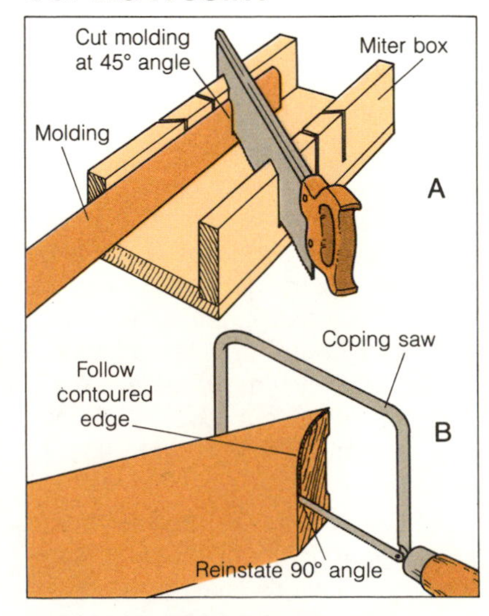

For a neat coped joint, first miter the end at a 45° angle (A), then follow the exposed edge with a coping saw (B).

Now measure for the side casing. If the door jambs are level and plumb, all should join snugly. If not, you'll have to adjust the angles of the side cuts to fit the top casing exactly. Then nail the side casings into position. If the finish floor is not yet in place, remember to leave room for it at the bottom of the casing.

Window casing. Most window units require interior trim around the opening. The standard treatment consists of top and side casings, a stool atop the finish sill, and a bottom casing—or apron—below the stool.

Begin by penciling a ⅛ or ¼-inch reveal just inside the side and head jambs; then measure the width of your casing. Adding ¾ inch to the casing's width, measure this distance out from each side jamb's reveal and make a mark on the wall. Now measure the distance between these marks: this is the length you'll cut the stool. Use either a flat piece of lumber or a preformed rabbeted or flat stool to match the slope of the finish sill.

Position the stool so that its ends line up with the marks on the wall; then mark the inside edge of each side jamb on the stool's back edge.

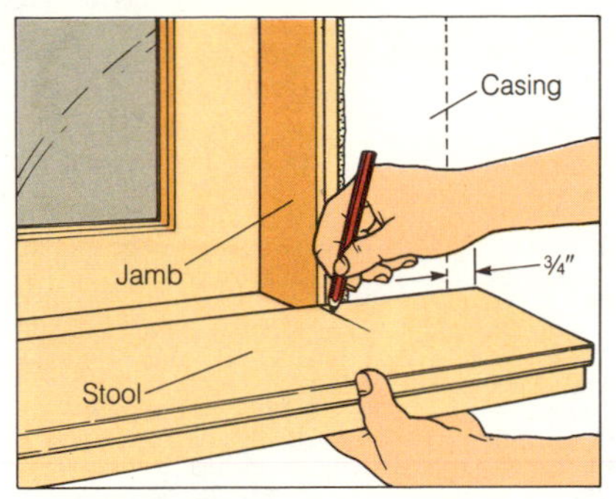

Mark the inside edge of each side jamb on the stool's back edge.

Place one end of the stool against a jamb (the back edge flush with the window's sash) and mark the jamb's front edge on the stool. Repeat this process for the other end.

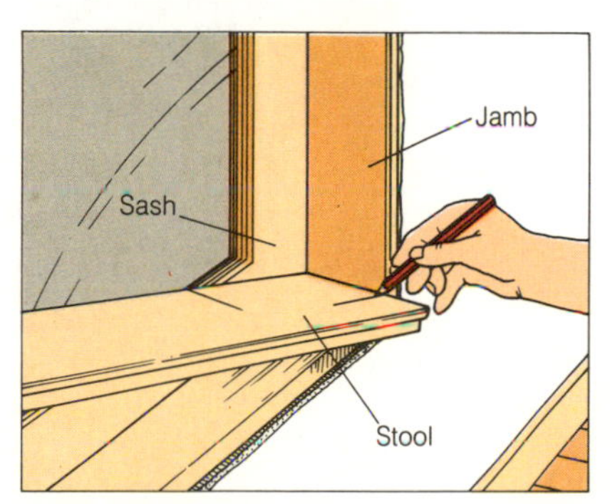

Next, mark the jamb's front edge on each end of the stool.

Using a combination square, extend each set of marks until they intersect; then notch the stool along the lines. Set the stool in place and fasten it to the finish sill with 6-penny finishing nails.

Now, square off one end of a piece of casing. Set that end on the stool, aligning the inside edge with the reveal. Mark the inside edge of the casing where the head jamb's reveal crosses it. If you're using contoured molding, cut the end at a 45° miter. For flat lumber, cut the end square.

Nail the casing to the jamb with 4 or 6-penny finishing nails, and to the rough framing with 8-penny nails. Repeat this process for the other side casing.

On another piece of molding, cut one end to fit the side casing. Make a trial fit, then cut it as necessary so the casing follows the reveal line. At the other end, make another trial fit; if all is well, cut the casing to length and nail it in place.

For the apron, cut a piece of molding to the length between the outsides of the side casing. Center the apron under the stool and nail it to the rough framing with 6-penny finishing nails.

INSTALLING DOOR CASING

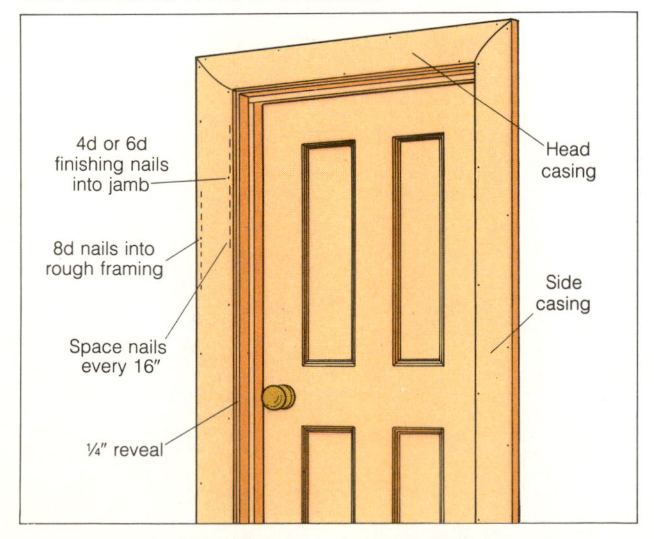

To trim a door, first pencil a reveal line on each door jamb; align casing edges with the reveal and nail casing to both jambs and rough framing.

WINDOW TRIM & CASING

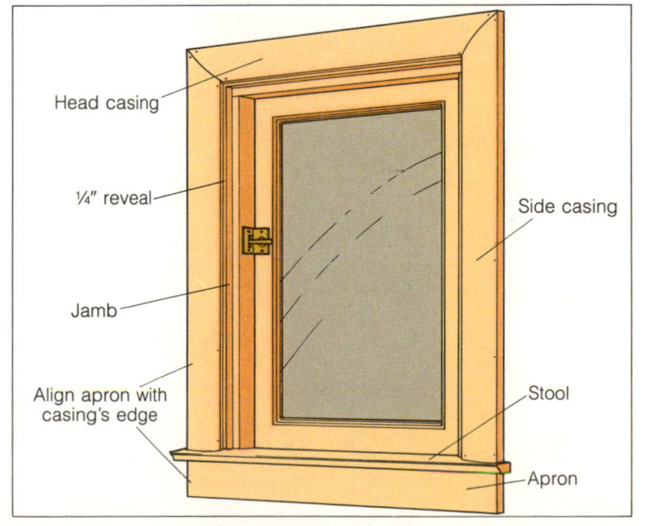

Finished window trim consists of casing, stool, and apron. The outside edges of casing and apron align; the stool extends slightly past this point.

Index

Sunset Proof-of-Purchase
ISBN 0-376-01015-0